DIRT

DIRT

a love story

EDITED BY
BARBARA RICHARDSON

FOREWORD BY
PAM HOUSTON

ForeEdge

ForeEdge

An imprint of University Press of New England

www.upne.com

© 2015 Barbara Richardson

Manufactured in the United States of America

Designed by Eric M. Brooks

Typeset in Fresco by Passumpsic Publishing

For permission to reproduce any of the material in this book,
contact Permissions, University Press of New England, One Court
Street, Suite 250, Lebanon NH 03766; or visit www.upne.com

Library of Congress Cataloging-in-Publication Data

Dirt: a love story / edited by Barbara Richardson.

pages   cm

ISBN 978-1-61168-900-6 (cloth : alk. paper) —

ISBN 978-1-61168-766-8 (pbk. : alk. paper) —

ISBN 978-1-61168-800-9 (ebook)

1. Soils — Miscellanea. 2. Soils — Anecdotes.

I. Richardson, Barbara K.

s598.D5   2015

631.4—dc23   2015003396

5 4 3 2 1

I'M A DIRT PERSON.

I TRUST THE DIRT.

I DON'T TRUST DIAMONDS AND GOLD.

✳

EARTHA KITT

CONTENTS

FOREWORD
Scratching the Surface

PAM HOUSTON

I live on 120 acres of dirt in a high mountain meadow in the Eastern San Juan Mountains in south central Colorado, near the headwaters of the Rio Grande. A woman named Dona Blair sold me these acres for 7 percent down and a signed copy of my first book, *Cowboys Are My Weakness*, because, she said, she liked the idea of me, and 7 percent down was all I had. I didn't have a job, either, or three pages of a new book to rub together. But my father was a hustler and he taught me to be a hustler and so for the next twenty years, I accepted every writing assignment and every teaching assignment that I was offered, and several more that I wasn't offered but had to go out and rustle up. I didn't sleep that much in those two decades, but I love what I do for a living, and I am not sure our thirties and forties are supposed to be for sleeping anyhow.

Dona Blair still spends her summers in Creede, in a cabin on the other side of town, and what she always said about me to Bertie the postmistress, and Jeff the hot dog man, and Brad at the hardware store is, "You know she makes those payments, *and on time.*"

I paid the ranch off last year—every dollar coming from writing or teaching—on July first, in the middle of the largest fire in south central Colorado's history. We were on standby to evacuate, my animals sent to Gunnison for safekeeping and a few favorite belongings in boxes in the 4Runner, and I drove under the sickly orange sky through the lung-searing smoke to Dona's house and gave her the final check. All the giant spruce trees her husband had carefully designed the house to fit among had orange flagging tied in the branches. These were the ones the forest service would cut down if the fire got too close.

I thought maybe because I am a fiction writer and have an overactive sense of narrative, the ranch would burn down on the day I finally owned it, but it didn't. The monsoon came, as it has every Fourth of July weekend, on schedule, even in the drought years. You can set your watch by it. The Texans arrive by the thousands in their campers on the afternoon of the third, and then the sky opens up and it starts to rain, and it pretty much doesn't stop, thank God, until the end of August.

But twenty-two years ago, when I became the caretaker of this 120 acres of dirt, the monsoon was only one of a thousand things I didn't know about. I didn't know, for example, what a week's worth of thirty-five below zero can do to an unprotected pipe, especially in a particularly dry winter, when there has not been enough snow to create a skirt of insulation around the base of the house. I didn't know that in other years, it would snow so much the split rail fences would go under and stay there for weeks or even months before they surfaced, and all of us, humans, livestock, and wild animals alike, would walk around on a kind of featureless moonscape, a dome of blue sky over an unbroken slab of white.

I didn't have the slightest understanding of what it meant to live at nine thousand feet above sea level, how quickly and completely the sun's ultraviolet rays eat through paint on the south side of everything: the house, the barn, the car. Nor did I know that at this elevation there is an exact right number of animals that can graze on 120 acres so that the pasture stays healthy, and that number turns out to be much smaller than you think.

Local wisdom would say eight, but in practice I have found it is more like six—unless some of those are short, like the minidonkeys, Simon and Isaac, and the sheep, Jordan, Queenie, and Natasha, in which case eight will work out all right. As long as it's not a drought year. In a drought year, I have to supplement with hay in May, June, and, depending on how much the pasture recovers during the monsoon, sometimes even September.

A few years into caring for this ranch, I understood that 200 bales of grass hay in the barn would get me through most winters, but that 250 is a safer bet in case the snow stays on the ground till May. It was a few more years before I understood the relative value of grasses as measured by the inside of a horse's stomach—blue grama, fescue, timothy, bear grass—and recognized that blue sage coming in is a sign of overgrazing and that no one will eat it. Not the mule, not the mule deer, not the elk, not the horses, not even the snowshoe hares. By the third year, I understood that four cords of wood would get me through most winters, but it was better to buy five unless I wanted to drag a cord through the snow around to the wood porch on the front of the house.

When this 120 acres got entrusted to me, I had no idea what was under the dirt: sandstone, feldspar, gypsum, granite, and traces of copper and amethyst, silver and gold. Also under the dirt is the homesteader, Robert Pinckley, who filed the original mineral claim and who walked upon it longer than anyone. He is buried in a family plot at the top of the hill overlooking the barn, along with a woman and a child, neither of whom made it through that first brutal winter of 1912, but only Robert's grave is marked with a proper stone. Upon closing, I signed a piece of paper that said their descendants could come in perpetuity, anytime they wanted, to tend the graves.

When I signed the papers on this 120 acres of dirt, I didn't know who all walked upon it: coyote, red fox, black bear, badger, mountain lion, mule deer, lynx. I didn't know that a bull moose would hunker down in the willows behind my barn each year for moose season, right on schedule, as if he had booked a room. I didn't know I would be able to sit at my kitchen table in the extreme quiet that is January and watch a herd of two hundred elk move silently through my pasture, leaving the safety of the national forest above me, to go down to the river and drink. Rocky Mountain bighorn, martin, otter, ringtail raccoon, marmot, rock marmot, prairie dog. And me, of course. I walk on this dirt every morning. My horses and donkeys roll in it, my chickens scratch in it, Mr. Kitty hunts upon it, the Icelandic sheep dig themselves little earthen beds in it, and William the wolfhound . . . maybe he loves it most of all.

On the lengthening list of all I didn't know about this dirt, add the names of the things that grow up out of it: mountain sorrel, western fringe gentian, alpine aster, rosy pussytoes, western monkshood, mountain bluebells, Queen Anne's lace. I didn't know I would learn to mark the calendar by the arrival of flowers, the delicate wild iris at the end of May, the spindly blue flax in mid-June, followed immediately by the columbine, which come in all colors, though blue is the color Coloradans love best. Silvery lupine and Indian paintbrush paint the pasture in early July, once the rain begins, followed by sky-rocket, penstemon, and those little blue harebells that move in the breeze and look as though they are filled with light.

I learned the hard way that growing vegetables at nine thousand feet is, at best, an exercise in tenacity. I learned that the soil is mod-erately good, once you dig all the rocks out, but it can snow or freeze hard any month of the year and it usually does at least once every thirty days. I learned you can forget about tomatoes, peppers, corn, melons, basil or any other herb except rosemary and that leafy greens will grow in July, but the cottontail rabbits are surprisingly clever and relentless. I learned that a decent summer's yield for all those hours of labor might be six potatoes, a handful of radishes, two salad's worth of lettuce, one tiny pumpkin and a dozen cherry tomatoes grown in a kitchen window pot that never even made it outside.

Sitting on the back porch with a pair of binoculars, I learned who flew over this 120 acres: osprey, eagle (bald and golden), peregrine falcon, red-tailed hawk. Steller's jay, nutcracker, magpie, raven. Nut-hatch, bushtit, hummingbird, wren. Cliff swallow, barn swallow, tree swallow—so many swallows and all of them wanting to build nests in the eaves of the house. I learned too late that swallows carry bed-bugs, and I learned soon after that the only way to kill bedbugs is to drag the mattress out onto the snow for a night or three or seven and freeze those prehistoric suckers out.

I learned that woodpeckers appear in great numbers after a fire (all that wood to drill) and that a bald eagle will come back to the same nest only every *other* year, for sanitary reasons (maybe they have bedbugs too!). Above all other avian residents, I would learn to love the belted kingfisher most, for his colorful head and bright eyes and high-collared turtleneck sweater. I would also develop a soft spot for the owls (barn, great horned, and snowy), and the way their call can take the loneliness right out of a solitary winter night.

The anthology you hold in your hands contains essays by writers, travelers, biologists, sculptors, green architects, terrestrial ecologists, geomorphologists, soil scientists, environmental economists, Sufi teachers, medicine women, farmers and the daughters and sons of farmers, and people who generally like to live close to the land. Some of them, like me, have a particular piece of dirt they love so much that just saying the names of the things that live in and on and around and because of that dirt makes them unaccountably happy.

In her essay "Dirt Princess," Julene Bair remembers being a baby, crawling in the dirt on her family's Kansas farm and feeling "its hard humps beneath my hands and knees, its stubborn solidity in my belly when I fell flat, or sprawled for pleasure." In "The Language of Clay," Roxanne Swentzell writes about how often, as a child, she would lick the walls of her adobe home, and she swears she could tell the difference between the taste of her neighbor's adobe house and her own, which was sweeter. In "The Great Beneath," Linda Hogan sings a song of lyric praise to the dirt the wind blows into her house, the dirt she scrubs from her horse's rump, and the manure that feeds the bushes where the butterflies return in spring.

Some of these writers—the scientists primarily—love all dirt, equally and unconditionally, and recommend that we do the same. In his essay "Dirty Business," David R. Montgomery warns that "every nation's most strategic resource is the thin layer of rotten rock, dead plants and animals, and living microorganisms that blanket the planet. Yet we treat this natural capital like dirt instead of the frontier between the dead world of geology and the living world of biology, the living foundation for life on land."

Several essays confirm something I have suspected all along, that those who get dirty with some regularity and a lot of enthusiasm are more likely to be healthy than those who don't. And though the essays are wonderfully diverse in both subject and stance, the spirit of the anthology might be summed up in this anonymous quotation that Kayann Short found in her yearly seed catalogue: "Humans—despite their artistic pretensions, their sophistication, and their many accomplishments—owe their existence to a six-inch layer of topsoil and the fact that it rains."

Erica Olsen provides us with a dictionary of dirt, asks us to pause and consider the beauty of the words *aeolian*, *loess*, and *midden*, and suggests how we might use them as metaphors to access our interior landscape, the emotions we hide even from ourselves. BK Loren's essay reminds us that whether or not we keep track of our dirt, our dirt is keeping track of us. She writes, "Dirt is everywhere and records everything, retelling your story, perhaps even eons after your death, in sediments pressed into history, pressed into time. There is nothing you do that escapes record. There is nothing that the earth will not record and read back to you and others."

Janisse Ray celebrates the termites and the earthworms that will turn the mountain of broken and discarded objects her parents have spent a lifetime accumulating back into dirt and dust. Peter Heller writes about building himself a house out of dirt, a house that captures and holds the heat from the stove and pulls in heat from the south-facing windows and therefore never freezes, a house that will one day return to the earth, as, he realizes, he will himself.

From dust thou comest, to dust thou shalt return, says the Bible, as Marilyn Krysl reminds us in her essay "Services at the Church of Dirt," which mixes up God and Earth and Mystery in a way that is most pleasing. When Marilyn was a child she spoke to dirt, understood that it had magical powers. She fashioned Adam and Eve figures out of clay, just to bury them in the yard and bear witness to the earth's profound reclamation. "Sitting in the dirt, without instruments," writes the grown up Marilyn, "we are sentient creatures. Sitting in the dirt, without instruments, we begin to see."

What I learned first, and best, and will continue to learn forever, from my 120 acres of dirt in a high mountain meadow, is that ground, like all the best things in life, gives itself up slowly, that in my twenty-two years here (and counting . . . always counting), I have only scratched the surface of all the things there are to learn.

With that in mind, I walk out the kitchen door each morning and down the dirt path to the corral, where the horses and donkeys are nickering because they know my pockets are full of apple slices and carrots. After they have turned my pockets inside out and gotten every last slice, I pull a bale of hay down off the pile and drag it out to the pasture with the hay hooks, break the strings and put them in my pocket, and then spread the bale around widely enough so that neither Roany nor Isaac can bully Deseo or Simon out of their fair share. It's late October, and the grass in the pasture is losing its nutrition and the farmer's almanac is predicting a big winter, and I have learned to get everyone fat and happy before the deep freeze sets in.

I listen, for a moment, to the deeply gratifying sound of four equines chomping hay in the silent dawn. I check their mineral lick, to make sure Isaac—our resident clown in donkey's clothing—hasn't turned one of the feeders over on top of it, and then walk the dirt path to the trough in the east pasture, the one that is fed by the frost-free hydrant. I double check the heater and top the big trough off, then fill a couple of buckets to carry to the sheep pen. I'll give the girls—Jordan, Queenie, and Natasha—a flake of hay and a promise to let them out later to eat the still-green grass around the house.

I won't have to call William to my side because he's there already, knowing it is time for the walk we call the *long pasture*. We start from the corral and head southeast, to the quarter section post where my ranch meets the Soward Ranch, and then follow the west fence line its entire length, the Rocky Mountain bluebirds who haven't left, yet, flitting along from post to post in front of us, as if we all have decided to take a walk together. The little creek that drains the wetland at the back of the pasture is running strong after last night's snow up on the mountain, so we find a narrow place to leap across it, and then climb the stile over the south fence and into the national forest.

We follow the dirt trail, past the place where, more than a decade ago, Rose the wolfhound came barreling out of the woods with an angry cow elk on her heels and I had to clap my hands three times and yell "Stop!" to keep from being trampled. We climb the hill past the little canyon where a character named Logchain, who fashioned knives out of old truck springs and wore only clothes made from leather he had tanned himself, hid out from the Missouri State Police for almost two years before they came to get him. We meet up with Red Mountain Creek and I wait while William takes a sip and leaps in after a few trout that were darting around to tease him. We follow the creek another five hundred yards along a carpet of orange and yellow aspen leaves until we get to Sally's bath, named for a coydog who lived with me on the ranch for seventeen years and who liked to roll in dead things. A traditional bath with a bucket, hose, and soap would send her under the porch sulking for days, but she didn't mind being chucked in to the deepest hole on Red Mountain Creek, and it reduced her odor at least by half.

From Sally's hole it is a steep but brief climb up to Coyote Rock, named one day a decade ago when a coyote flirted with Rose up there for more than an hour before we got a little scared for both of them and ran the coyote off. That is our turnaround point, and we make our way back to the ranch, back over the stile, and around the pasture the other way, through the wetland, where a pair of mergansers have been making their home all summer, and past the gravesite, which I never pass without saying thanks. When we make the final turn for the house, William stops and smiles back at me. I like to imagine that, just like me, he can feel the goodness of this ground coming right up through the dirt and into his feet.

PREFACE

The God of Dirt

*

BARBARA RICHARDSON

For thousands of years, humans have looked to the heavens for inspiration and divinity. Looking to the heavens may be the greatest mistake we, as humans, have ever made. We project what we want onto the open skies, the blank distant blue. Whereas looking to the earth sends clear messages—intricacy, impermanence, solidity, interrelation, humility. You can't fool dirt. Nor can you escape it. You can't manipulate meaning as you can from the mirror of an empty sky. Dirt anchors us all in reality. And so we need to remember and relearn the ongoing, resonating divinity of dirt. As John Keats wrote, "The poetry of the earth is never dead."

That poetry is everywhere. It comes in through all our senses. Green, gold, scaled, seeded, sour, shining, sneaky, squeaky, voluminous. . . . Mary Oliver writes, "The god of dirt / came up to me many times and said / so many wise and delectable things, I lay / on the grass listening." The essays in *Dirt: A Love Story* are that listening. Remember the joyful freedom of splashing in a mud puddle? The thrill of climbing an eroded cliff? The artists, scientists, and authors in *Dirt: A Love Story* drag you outdoors, scuff your knuckles, and muddy your feet. They make dirt live and breathe again.

The first set of essays, "Land Centered," returns dirt to its rightful place—as the crux of life in the experiences of people who are flagrant dirt fanatics. These writers revel in the fact that dirt is "magnificently humble." Long may they reign. Then, armed with new appreciation, take a muddy fall into "Kid Stuff," the second set of essays, which explores our early contact with dirt. Go ahead, these writers say, "major in mud pies." Because the humbling, hallowed fact is that dirt is our mother. And she doesn't call us inside at night in order to ignore her gifts.

"Dirt Worship," the next set of essays, shows just how to get "that motherly feeling" on in adulthood. How to place your feet on the ground and your hands in the soil and claim your ancestry, your grand, mysterious inheritance. This centering in the land leads to curiosity about the good stuff under our feet. And so the fourth set of essays, "Dirt Facts," offers insights into the masterful and largely ignored scientific processes within dirt, the "interesting secrets" that children and dogs may not understand scientifically but enjoy with all their hearts.

Lastly, the essays in "Native Soil" embrace the challenge of adoring seemingly unlovable ground — third-growth woods, weedy urban lots, rock-hard prairies — "the sort of land that desperately needs to be loved and protected, and rarely is." These essays salute and defend our native soils as if they were life itself, which they assuredly are. "Humble" comes from *humus*, ground, and *humilis*, lowly. Humble outdistances pride. Humble whispers connective language, and it waits when we don't listen. By book's end, you will recall the generous, wordless, irresistible divinity of dirt.

That divinity says get filthy. Grab a shovel. Hike a ravine. Breathe a dust storm. Reek like old goat and sleep like Venus after a dirty long day. Relish dirt's unbiased receptivity. Worship, if you will, the endless fecundity of soil. Or better yet, fall in love. Dirt makes a resilient, astounding lover. Tireless. Generous. Unstoppable. And most often unthanked. Start thanking. Put your belly on the ground and say *thank you*. Wherever you are. Winter, spring, any season will do. Lie there saying *thank you* until all of your internal chatter and sophisticated notions and cogitative claptrap stop.

While you're down there, imagine every plant that has ever lived. Every seed that has dropped, every band of people, every fish in every stream, every hedgehog, every grasshopper, all the grasses of all the prairies on earth are still here. The trees. The elephants. Every single ant and albatross. You needn't try to imagine it; it is so. Under your belly. The earth should be groaning under piles of its own dead life forms, but what a spacious, cleanly earth it is. Right beneath you lies a creative silence so vast it makes time stop.

Walt Whitman, long gone from us, said, "Look for me under your boot-soles." He meant it literally. This astonishing vanishing act to which we belong deserves consideration. And deep respect. Respect for the arbiter of this vast balanced nuanced productivity. Let God in heaven take care of the stars. We, along with the scientists, artists, and poets, are forever called to loving dirt.

1

LAND CENTERED

magnificently humble

MY HOUSE IS THE RED EARTH;
IT COULD BE THE CENTER OF THE WORLD.
I'VE HEARD NEW YORK, PARIS, OR TOKYO CALLED
THE CENTER OF THE WORLD, BUT I SAY
IT IS MAGNIFICENTLY HUMBLE.

JOY HARJO
SECRETS FROM THE CENTER OF THE WORLD

MY LIFE IN DIRT

❋

EDWARD KANZE

I was born dirty and I'll die dirty. Upon my delivery, slimy and blood-streaked from an amniotic sac into a hospital room, a doctor pronounced me alive, a nurse wiped and swaddled me, and my mother began to supply the milk of maternal kindness. If I die in a hospital or at home, I expect to be zipped into a giant plastic bag, as I saw done with Pop, my ninety-year-old father's father, and whisked off to a mortuary. There someone else will clean and swaddle me one last time.

Between the coming and the going, we live, work, and play in a dirty world. There's no getting away from dirt. Who would want to? Dirt is us, even though it sounds ungrammatical to say so.

As a child, pretty early on I learned there was a direct connection between dirt and fun. If I played indoors at a friend's house or outdoors on a ball field, I tended to return home pleased. But if by the time I parked my bike in the garage or stomped through the front door I was stained by dirt from hair to sneakers, words such as "euphoric" and "transcendent" measured my exaltation. The connection became clear. To live life full bore, you have to get really, really dirty.

Life's great pleasures—running around in woods, romping in fields, making love, raising children, traveling to enthralling places, helping those in need, immersing hands and spirit in sun-warmed garden soil—bring us into intimate contact with dirt of one kind or another. I have no doubt that dirt, as long as we keep it out of puncture wounds, is good for us. In recent years epidemiological studies have shown that exposure to a reasonable amount of good, old-fashioned dirt is a key to preventing the development of allergies. Cleanliness is not next to godliness. After a point, it's foolishness. As my kids would put it, being clean all the time is stupid. Bring on the dirt.

It turns out that the dirtiest of dirt, that which comes from our bowels, can be the magic bullet that shoots down someone else's gastrointestinal complaints. Where once we obsessed over hygiene around toilets, now we talk of the miraculous efficacy of fecal transplants in curing chronic ailments. Who could have imagined it?

I could have. Call me a dirt-o-phile. I'm drawn to dirt and always have been. I write this sentence on my fifty-seventh birthday. If there's one thing life has taught me so far, it's this: where lies dirt lies fun and adventure.

The only time in my life where I was offered a chance to get really, truly grubby and turned away came in Big Bend National Park, in the Chisos Mountains of far western Texas. I was feeling blue. My worries were the usual ones for writers. Where would I earn money to pay for the next bag of groceries? Was I inching far out on a professional limb, one that might break rather than grow? We were camping. My wife was shampooing her hair in the bushes. Wandering off, I heard soft grunts around me. I realized that without meaning to, I'd fallen in with a herd of wild, pig-like animals called javelina. There were fifteen of them, maybe more.

For what turned out to be a spirit-lifting hour, I roamed with the animals, talking to them and taking appreciative note of their banter. If I got a little too close, I'd see fur spike up on their backs. So I'd retreat. The javelina seemed to forget about me after a while. Eventually they picked their way down a rocky hillside to the shadowy bottom of an arroyo. Here there was a mud hole a little smaller than the average bathtub.

One by one, the javelina took brief, contented rolls in the mud. When my turn came I was tempted to follow. Indeed I would have had I not seen animal after animal spray urine in the wallow. Perhaps that's what kept it moist. Suddenly I thought of better things to do.

Geophagy, the eating of dirt, is usually considered a serious psychological disorder. Yet there is some evidence that a little dirt, if it's not tainted by toxins, human fecal matter, or parasites, may be good for us. Just as a lack of exposure to the world's marvelous diversity of dirt contributes to developing allergies, so an excess of cleanliness in our diet may have something to do with the GI tract complaints that are rampant in modern industrial society.

Shortly after I graduated from college, I took a position as a naturalist in a privately operated nature preserve. Among the items on my job description was caring for injured and orphaned animals. I soon found myself raising a brood of robins fresh out of the egg. Someone in the know told me to feed the baby birds cat food. This I did. The ugly little things gorged. Yet something was amiss. The birds developed chronic diarrhea. I consulted with a veteran naturalist named Kaye Anderson. Kaye told me to feed the robins dirt. "The cat food's OK," she said. "But also give them worms, dirty ones. To develop healthy digestive systems, the birds need microorganisms that occur in dirt. If they don't get them, they won't be healthy." I did, and the robins flourished. For them, dirt proved just what the doctor ordered.

Never have hands been more dirty than when I used to go fishing with my mother's father. No prissy dry flies for Grampy. He was a hook and worm man. Every summer he'd take me fishing for a week or two. We'd set off with a car and trailer (later with a school bus he converted into a motor home) with an ice chest filled with fresh moss. In the moss wriggled hundreds of live night crawlers Grampy had purchased for a few cents apiece from local kids. The kids plucked them from their yards at night, and Grampy took pleasure in giving these entrepreneurs spending money.

If you've ever fished with worms all day without access to warm water and soap, you've experienced grunge at its most supreme. Night crawlers exude slime as much as they crank out fecal matter, which, of course, is just dirt in the making. The slime serves as the glue that fixes the excrement to your hands. Because you're sticky, you also pick up every speck of filth you touch. By lunchtime you're caked with the stuff. Without a good hard hand washing, there's no getting most of it off. You eat anyhow, supplementing your intestinal flora just like the robins.

I learned many things from my grandfather. One was to take dirt in stride. He concerned himself with the important things in life — family, friends, outdoor adventure, treating everyone with respect and forbearance — and let soil, slime, and fecal matter fall where they may.

You know what? Those days on lakes and rivers, wetting lines, catching fish, getting caked in dirt and worm shit, were among the happiest of my life. It was the kind of happiness that didn't bubble to the surface and make a show erupting. Rather it pooled in a secret place in the heart. No one could see it, at least not directly. Yet it was there. Forty years and more later, the joy remains, a reservoir I tap during the hard times.

Thank heaven for dirt—literally. What is it, after all, but stardust? Some of it sifted just last night out of the cold, black intergalactic void. The rest fell from the cosmos last week, last century, last millennium, or a billion years ago. Vintage dirt has covered ground. Organisms of one kind or another gathered and ingested it and built the useful elements they gleaned from it into living tissue. Then came death, but the dirt played on. Liberated once more, these fragments of asteroid, comet, and supernova circulated anew and were drawn into new lives, again and again, on and on and on, from the beginnings of biological time until the present day. From here to eternity, or at least the biological equivalent of eternity, the dirt will rock on, continuing to bring grace to those who borrow it awhile.

It stretches the mind to think where dirt comes from and where it goes and how it moves around. In moving and distributing dirt, wind plays a major role. The famously fertile loess soils of the Mississippi River watershed, China, the Ukraine, and elsewhere consist of fine particles of sand, silt, and clay that were carried into the air largely during interglacial periods and then heaped up in locations where farming thrives today. The word "loess" comes from the same root as "loose." I was taught to pronounce it "lerse" by a favorite college geography professor. Truth is, there are almost as many ways to pronounce "loess" as there are different kinds of the rich, granular stuff.

Dirt moves in water, too. Hence the nickname "Big Muddy" for the sediment-rich Mississippi. Sometimes dirt moves as a viscous fluid, inspiring scientists to invent marvelous words such as "solifluction" to describe its thick and oozy flow. Glaciers move dirt, too, scraping the stuff from one place and heaping it in another. Glacial deposits called moraines include dirt along with a mixed assortment of grit, pebbles, and boulders. Truth is, dirt does not stay put for long.

No matter where dirt comes from and where it goes, and no matter what you call it, this humble material is more valuable in the final analysis than diamonds or gold. You can't eat precious stones or rare metals, but you can eat dirt, either directly as a geophage does or indirectly in the form of ingested organisms that are essentially animated, self-replicating forms of soil. We are what we eat, as the saying goes, and what we eat, in a roundabout way for the most part, is soil.

What a pity, then, that in our overly tidy culture dirt has taken on pejorative meanings. We talk of "dishing the dirt" in censorious tones, and of sexually charged movies and stories as "dirty" or "filthy." We pester children to clean their rooms, clean their bodies, and clean their minds.

Keeping clean isn't a bad thing, embraced in moderation. Still, it's a dirty world. To enjoy it, to engage with it and our fellows in enterprises worth the doing, we need to wear our dirt like a badge of honor and not be cowed by the obsessively, self-righteously clean. I love getting dirty. Just as Grampy did. There's plenty of time to be clean after we're dead.

THE GREAT BENEATH

*

LINDA HOGAN

*D*irt. I love to have my hands in earth. I can't pass by the plants I care for on any given day or season without caring for them in some way. Perhaps I move earth about for water to be better taken in. Or I pile more autumn leaves around them as compost and mulch. A Neruda poem contains the line: *What is a man without having his hands in the earth?* The same goes for a woman.

Besides, I live with it, with dirt, in a tiny cabin that is an island in the center of mountain wilderness. Here, trees are not in rows, but in their own intelligent chaos of growth. All the roads toward my place are dirt roads and dust floats through. The cabin itself is in a surrounding bowl of great mountains, trees, and a creek snaking just below. The one dirt road that arrives here is at the end of its destination, having come down a hill, often muddy.

It was the land that mattered to me more than the little house built back in the 1930s for summering over. It had roofing shingles as siding. It was falling apart, but the outdoors was an extension of the house and it was my living room, my world. I knew it from the beginning, twenty years before it was for sale. The mountain world manages to bring all its furnishings inside, dry grass, dust, pebbles. It enters in all ways, feeling at home here. It is never what anyone could call clean.

The cabin was built by one with a love for the land, built *of* the land, its stones the foundation, the fireplace made from a vein of quartz stones, a pink-white matrix in near earth.

Here, during the day, the mustang and burro roll on the ground with great enthusiasm, their bellies revealed in sudden, unexpected vulnerability, legs bent upward, and with sheer muscular will, they work at turning themselves completely over in their special places of dust.

Once they succeed in a complete roll, they rise up and shake off a cloud of soft dirt that will later be on my hands and under my nails. It would never have occurred to the young burro to do such a thing as roll over this way. In truth, I don't think she was made for it. But she imitates the horse, even trying to make some of the same sounds.

Each morning I brush them. The dust flies off, bits of dry plants come out of hidden places in manes and fur. When the coats are smoothed, I scratch their rumps, an action the burro especially loves. With the mustang, I clean the dirt from between her nipples and the smell of it is strong on my hands, an odor like herbs and the dank sweetness of her body. She smells nothing like horses bred by people for certain special qualities. As for the burro, those great long ears, how she yearns to have them rubbed and cleaned, how soft they are.

Then it is time for the shoveling of manure, although I leave some of it for the numerous crows, magpies, and other birds that flock down to pick through. They come in such large numbers, the crows, and if startled, they rise like one dark blanket lifted by a wind. Manure is a precious thing, those droppings with enough moisture to attract honeybees and butterflies in their season of spring. Bushes grow in this area. I call them butterfly bushes because they sometimes become completely white with the insects that love the blossoms and also love the manure.

For me, there is no escape from dirt. Even after I come inside, it comes with me. I step in and remove my work boots, leaving the earth or mud on the rug at the door, then I change into something clean, wash, and brush my hands and nails. But dust from our little traffic on the road above and from the fields sneaks in. It filters in from air itself, and settles on everything, on shelves, furniture, my collection of small, carved animals, my saved stones on bookshelves, and on the floor, which needs constant daily sweeping.

I could begin my love of dirt with memories of being a child dragging a magnet through the ground and discovering the iron filings standing straight up at the ends of the magnet. Or entertaining myself on the ground, making muddy clay into foods and dishes, making pathways that served as imaginary roads, breaking open rocks with a hammer on the unplanted ground. Even my father's forbearance when I decided to find China on the other side and dug until finally I grew tired of searching for other worlds. I didn't know that the earth I moved was a country with so many inhabitants already there and alive, and I was disturbing it.

As an adult, in my smaller carelessness of planting, I have shoveled land, used a hoe, only to have the intelligent bindweed appear to keep the earth from eroding, a miraculous plant stored hidden beneath for an unknown amount of time, awaiting destruction to rise. It is the same way cedars rise up in my homeland of Oklahoma. They are there to keep the earth cool, to assist the hardwoods, but are for now considered an enemy and cut or poisoned. No greater mistake could be made for the earth, and that tree might have saved it from the drought of late years.

Now I might also consider the winter snow outside, watching it fall, knowing that a bit of dust is at the center of each flake, that this dust went to live in clouds and that miraculous shapes formed around it. The snow, I now consider, will seep down into the ground and the waters beneath and travel through rocks and mysterious paths in the darkness beneath us.

One night, working outside in one of my planting places, I heard sounds on the earth nearby. Turning a light on it to see what was making the noises, I was surprised to find the ground was alive and swarming with earthworms. They wove over and around one another, moving about, tumbling on top of the dark, enriched soil. It is a mystery I still contemplate. Certain there was meaning to this, I wondered if the earth had shifted beneath the ground, or if something strange had come about inside the land, a minor earthquake, any possibility. That was many years ago. I'll never know what brought so many lives to the surface to churn and roil in such ways, how so much sound was created, because it seemed more than just the stirring movements they made, but as if the earth itself was talking.

Earthworms have earned my respect. I try to feed and grow them where they live in the ground. When I find them underneath manure, I am careful to move them to my gardens where they enrich the soil. Dirt.

They are the only keystone species I can name that are an invasive species. They arrived here from Europe along with their imported peach trees, although other lives flourished in the soil to keep it alive and rich. Now the worms are great creators and earth movers.

Darwin was fascinated by earthworms. His writing on soil was a bestselling book in its time, but the worms were what he mostly focused on in his later years. While they are a blessing to gardens, they are detrimental to hardwood forests, yet in slow journeys they have traveled to some. Those forests they reach fall into an unhealthy state of being. The presence of the worms destroys the natural growth of forest life. They take in and consume the fallen leaves and the other elements of forest creation that feed saplings, ferns, and many of our medicinal plants. We once had the wisdom to keep smaller plants and trees in their own places, growing in levels, the forest tendering down to smaller plants, berries, then medicines, then to foods and the flowers that repelled insects. Cool burning was practiced in most locations.

What we call dirt has a mind of its own, many minds, and it is filled with great lives and mystery. Loud and noisy worms are only one part of it. Inside the ground beneath us is a community of lives, mostly mysterious to us. In the layers down and down are lives as unknown as those in the depths of the ocean. In that dark community of earth is a balance we have been at work destroying, thinking only of the surface. At vibrant odds with what we believe we know dwell amazing soil and biota, fungi, the fine bacteria trees call toward them when in need. Even trees communicate with one another in the great beneath.

The layers of earth, the source of everything we think we know, have an unseen action of moving rocks, insects closed in winter, rodents and moles mistakenly believed harmful. We have snakes here, but much deeper are secret lives, and the greatness of it all reminds me that each step we take is walking not just on dirt but on incredible forms of life. In my own brief tenure here, this earth has changed drastically, some of it the natural processes at work, but some of it is devastation by human activity.

This year, in this one place alone, two freak storms arrived suddenly. The first was a crashing rain and great hailstorm. It washed down the mountain behind my home and it landed around my road and cabin, carrying rocks so large I couldn't move them from where they landed. Then the flood. The elder who helped build the cabin told me if it flooded to get away quickly. At the time, I thought *things have changed. Water doesn't run as it did in older days.* But great trees were felled by floodwaters, earth washed away from the roots of others. Masses of pine needles and other debris piled up against anything that remained standing. The small creek that runs slightly below this cabin became a river ten feet deep and raging through, taking with it what it could. It carried everything in its path. Other people came to watch the sudden surprising power as water carried great trees with it, crashing through muddy waters, and uprooting other enormous trees below the house. Trees I loved now lie down on the hill. They are bare of skin, their torn roots revealed to the sky. When I first went to the familiar places I walk, I found nothing the same. Natural pathways first made by animals were gone. In one place where water changed land, it had turned into a divided creek, as if another spring had risen from a secret place beneath. The mud came into the downstairs of my home, now cleaned more than once, leaving me unable to return things to their places. The horse and burro shelter is also somewhat undone.

But the land itself is chaos, and beneath the land that chaos has its own kind of order, missed by us because we are not thinkers of deep earth and what lives there.

Living in the world above ground for now, not yet ash scattered where I desire, I live surrounded by a weave and jumble of branches, trees whose presence I want to be near, and when I see them, I think of the unseen systems beneath, even greater and more interwoven than the latticework above where right now the barn swallows fly back and forth, carrying mud to build their beautiful clay houses, birds living in their own earth pottery creations, homes to birth and house their young. I watch them return home all at the same time each evening, quickly entering their own dwellings. And the creations of mud daubers are castle-like towers of earth, lengths of layered and powerfully strong columns for their young to grow in throughout winter. These look much like stone shaped down to finery by wind in our red rock deserts. Made of earth, clay, water, even spit. What we call *dirt*.

I study the great history of this land, once an inland sea, later the place where dinosaurs walked, then the indigenous people passing, camping here by the water, and it reminds me of my own people in a strong and once sustainable past. We want our world to be sustainable now, and for the future we have to remember that it all begins with soil.

The trees we need rise from it, as do the medicines, which may look like weeds. In the most damaged land of drought-ridden Africa, the planting of trees changed even the earth soils, attracted animal life, called back old birds, and drew to them the waters. Wangari Maathai received a Nobel Prize for this planting of trees. She began the Green Belt Movement. It returned life to her starving, drought-cursed people, merely starting with the planting of trees. Even now the planting continues. It brings back all that had disappeared and been used up by those who didn't know that dirt might become the beginning of life.

Here, meanwhile, the burrowing creatures and owls live in the ground nearby, as do great snakes in their dens, snakes who return to the same place for hundreds of generations. All these intelligences dwell in the beneath. Many are lives not often seen or known, and those that are visible are sometimes feared or unwanted, even the spiders who so carefully build their worlds and create entrances of woven dry grasses or trap doors of intelligent design. They are not threats, but significant other lives in our world. The world they live in is alive and too often it goes without our appreciation, yet all lives in that darkness are giving us something so generous that even if we don't understand it, we need to respect it, or thank the glorious dirt and all its inhabitants. Dirt is their element, even with the horse and burro running over it, hooves pounding it down.

I think of God ordering Moses to remove his sandals because he stood on holy ground. It is all holy ground. We should sing to it, all of it.

DIRT FANTASIES

❋

JANA RICHMAN

I dream about digging in dirt. In my fertile imagination dirt begins at sensuality, climbs the trellis of eroticism, and drops into
the hole of debauchery, where it romps lasciviously before climaxing
in rebellious abandon.

Gorgeous, sexy people dig in dirt. People who age well. People who
collect beauty in the creases of crow's feet. People with sturdy hands
and good minds. In the evenings, dirt diggers dine at friend-encircled
tables, where laughter and wine pour forth in equal measure, where
confident, unjeweled fingers twirl glasses, where dirt persists under
nails and in cuticles. The diggers taste their food with more intensity,
more essence than the nondiggers. The word *luscious* comes to mind.

I've been digging in dirt all day, they say, and then, *I love digging in
dirt*. The words enter the atmosphere with such lustiness you quiver
with images of black, loose dirt fondling your own fingers. And the
dirt . . . oh, the dirt! So perfectly textured, like silk drawn over erect
nipples. The dirt collects heat from the sun and offers it to you. But
then, as you reach for it, the dirt playfully tugs you into its cool re-
cesses. You're surprised when a barely audible moan spills from
your lips.

Dirt fantasies, like sexual fantasies, vibrate with tension between
fantasy and reality. Both are genuine, and one plays off the other, gently correcting and merging the romantic with the banal, the wild with
the tame, the unconscious connection to dirt with the conscious experience of dirt.

The dirt surrounding my desert home is not perfectly textured. It
is not rich and dark and moist and soft and cool to the touch. It does
not gently stroke my arthritic fingers. It is not being aerated by slow-
moving, benign worms. My dirt is dry, hot, hard-packed clay occupied
by copious members of the family Formicidae—the industrious ant

—which explore the giant in their midst by biting, ascertaining, I assume, my suitability as a food source and the possibility of carrying me away crumb by crumb. I try to cultivate a gentle attitude toward ants because I'm a fan of E. O. Wilson. I know ants are busy cultivating my clay-like earth where worms fear to tread, but instead of dropping to my knees, lowering my face to the ground, and offering thanks, I swear and stomp at them. "Stop biting me, you little bastards!" I scream.

Ants are dirt fantasy killers. Earwigs are too. Flies also. Along with whatever gopher-type animal leaves loose dirt mounds—an impressive feat—on the surface of my small plot of solid earth. All these busy creatures share my dirt and outnumber me by maybe a million to one. They also share the products of my dirt, which is, I suppose, only fair. The legal authority proclaiming the dirt mine doesn't hold much sway with them.

Although my dirt is not the dirt of my dreams, it has taught me this: dirt does not need to achieve my idea of perfection to produce. Out of my broken clods of clay sprout tomatoes, squash, corn, peppers, and whatever else I manage to shove into the ground. Apples, pears, cherries, peaches, apricots, and plums drop onto concrete-like earth in my backyard year after year. Greenery breaks through also—a few blades of grass but mostly weeds that we mow and call "green space." I've been told I need to bring in "good soil" and work it into my dirt. That would certainly feed my fantasy, but there's something about the toughness of my dirt and its ability to create new life in its current state that keeps me from doing so. It seems a betrayal, like the sixty-year-old husband replacing his sixty-year-old wife with a younger woman, one smoother to the touch.

My vision of myself as a dirt digger is akin to that of the "avid hiker" with the never-worn, expensive Italian hiking boots in a dark corner of the closet. Gardening is not my strong suit. I cram vegetable plants into my dirt—invariably too close to one another—and let them fend for themselves among the weeds. And they do. They hold strong. My garden is not suitable for the pages of a glossy magazine—it is an entangled mess that requires bravery and a machete to harvest tomatoes—but it is hardy.

Still, my dirt fantasies remain intact. They are not frivolous. They are not based solely in romanticism, but I've come to realize that I'm not so much a dirt digger as I am a dirt wallower. Wet or dry, I love dirt on my skin.

My first recognition of this came in the summer of 1962, before I started first grade. A friend and I were playing, as was our custom, along the ditch banks behind my house. The dirt of my current reality was also the dirt of my youth, so the ditch banks were made up of soaked clay-like earth. We had never heard of women going to spas for mud treatments—we had never heard of spas—but covering ourselves in cool, gloppy mud seemed an irresistible idea. It was also an idea that would land us in a heap of trouble, and we knew it. So we compromised. Next door, a younger boy had set up a classroom of stuffed animals and was in the process of dressing down a giant giraffe that had apparently spoken out of turn. Because the boy swiftly determined we were having more fun than he was, we made quick work of convincing him to sacrifice the giraffe to our cause. We would slop mud on the long-necked beast and cool ourselves in the process. The brilliant idea generated hours of reckless, muddy fun. Why we thought this would lead to less trouble is no longer part of my memory. As we would find out from the boy's red-faced, furious mother, giant stuffed giraffes don't grow on trees, are nonwashable, and are meant to live indoors. Even though the giraffe was outdoors before its spa treatment began, that technicality didn't save us.

In his book, *Magical Child*, Joseph Chilton Pearce identifies the living earth as the second bonding matrix—after the mother—in a child's intellectual development. A child has no capacity for abstract explanations of the world, the kind parents love to impart; her development is 100 percent experiential: mud feels good, doesn't taste good. Lesson learned. What else offers unqualified practicality but the natural world? If a child is unable to process the natural world through the body, says Pearce—rolling in sand and grass, eating dirt, chewing on sticks, sniffing flowers and dung, hearing the buzz of insects and birdsong—the patterns for practical sensory organization never form in that child's brain, and the creative logic of that child is forever thwarted.

My mother's purpose for sending me outdoors to play had little to do with her desire to create a magical child and more to do with her desire for silence, but in spite of her intentions, I found my place along the ditch banks and in the mud sloughs of the world.

I now leave stuffed animals out of my filthy activities—to this day I have disdain for their uselessness—but the urge to cover my skin in dirt has never left me. Desert quicksand after a monsoon provides a fleeting fix, but one can't typically get more than shin deep in it. A few years ago, after hearing about a natural hot spring/mud bath in a meadow near a small California town, my husband Steve and I drove six hours to sink ourselves into silky, black, sulfur-stinking mud. I can't imagine anyone not wanting the experience, but when I tell friends about it, they crinkle their noses and seldom ask for directions.

On my fiftieth birthday, Steve and I hiked to Boucher Creek at the west end of the Grand Canyon. The trail was harsh, the packs were heavy with extra water, and the temperature hovered around 110 degrees even though we had set out at 4 a.m. In the ranger's office the day before, we had been appraised under doubtful eyebrows, received a reluctant nod of approval, and were told we were on our own. Boucher Creek was difficult to reach, no other campers would be there, and no rangers would be coming to check on us. In other words, perfect conditions. The ranger warned us to be off the Tonto Plateau—referred to as the "death zone"—before 10 a.m., and we took his advice seriously. At 10:15, we dropped into Boucher Creek hot, dehydrated, and exhausted, stripped our bodies of packs and clothing, and lay on our backs on a flat rock in the creek, the tops of our heads used as a sort of stop log to divert water around us.

For the next few days, we lived as close to the earth as modern humans from an unnatural civilization can live. We wore only a pair of sandals, sprawled in dirt, rinsed under waterfalls, swam in pools, and dried out on rocks. Unless submerged in water, we were never without dirt on our skin. We pressed our bodies together often, finding pleasure in the grit and heat between us. Four days later, when it came time to hike out, we reclothed ourselves in the items we had worn in. Nylon shorts and lightweight, sweat-wicking shirts felt heavy and restrictive — even silly. We spoke little on the hike out, silenced by the sanctity of the experience and the sadness of its rarity. I felt as if I came out of Boucher Creek in a stronger body, but it had nothing to do with physical strength. A more accurate description: I came out of Boucher Creek more strongly embodied.

Since that trip, we have sought such encounters. Sometimes for days, sometimes for only an afternoon, we shed our so-called protective layer and put body and earth together. A spiritual retreat within a natural retreat center. The longer the exposure to dirt, the more firmly embodied one becomes.

It is not unusual for humans to feel an impulse to shed clothing in the desert. It might be one of the most common surges of animal instinct remaining in us. In his book *The Man Who Walked Through Time*, Colin Fletcher writes about experiencing a heightened awareness upon removing his clothing, a more intimate consciousness of the interconnected web of life in the Grand Canyon. Without clothing he felt more physically a part of the interwoven ecosystem, and he more felt more deeply the diminutiveness of the human time scale. I'm never surprised when I come upon naked people in the desert; I'm surprised it doesn't happen more frequently.

Without having conscious awareness of it, which is as it should be according to Pearce, I came out of my childhood with a profound bond to dirt and carried that relationship into my adolescence. There I found the charcoal-gray dirt of the Oquirrh Mountains near my childhood home — hard-packed with an inch or two of loosely floating topsoil that produced puffs of smoky dust when walked on. The barefoot hippie symbol of freedom reached my small town in the late '6os, a fad I enthusiastically embraced, partly because it mortified my mother and angered my father, but mostly because the human foot seemed then — and still seems now — a well-suited instrument for walking on the earth, especially those parts of earth not covered by pavement and concrete. In short, it felt good. I still don't understand why the bare foot is repulsive to so many, why it's okay to track dirt into a store or restaurant on Vibram soles but not on the sole of the human foot.

A vision of my feet covered in the silt of the Oquirrhs is one of the most vivid memories I carry, and that memory is solidly attached to *me.* By that, I mean, there has always been a small place in me that yearns for my own life—the life I want to live versus the life I'm supposed to live. The supposed-to life fits neatly into the expectations of family and society; the want-to life departs drastically from that picture. The two can never fully merge because the want-to life includes a small house and the supposed-to life demands that I pay for it. But my interior place of craving remains, and it is solidly connected to the earth's surface. Somehow, putting the human body in direct contact with the earth infuses it with unrestrained imagination. I've carried this knowledge with me since childhood, albeit mostly at an unconscious level. Once I moved it into consciousness, I've stayed close to dirt and moved steadily toward the want-to life, disappointing many along the way, but pleasing the wistful girl sitting on a creek bank staring at her dirt-covered feet.

The older I get, the more essential dirt wallowing becomes. It reminds me that I can age, my stomach can lose its tautness, my skin can loosen, my joints can stiffen, and my body can begin to wear out as it is designed to do, but until it drops, it is still my body. It is the same body that forty years ago enjoyed a pair of Levi 501s drooping on its bare hips below its flat belly; the same body that has forever objected to the idea of being strapped into *women's wear*; the same body that soaked up too much sun, turned scarlet, and shed its outer layer of skin; and the same body that kicked up puffs of gray dust around bulging tree roots in the Oquirrh Mountains.

It is the dirt on the body that reminds it of its sensuality, that allows it to claim its carnal appetite at any age. In a culture where only young women are allowed to express themselves as sexual beings, dirt on the body allows the old woman to say: fuck that. You don't determine my sexuality. I do. The earth does.

PRAISE TO THE TRANSFORMERS

✽

JANISSE RAY

I dread the job, but the room has to be emptied. Both of my father's knees are bad and the day will come, sooner than later, when he can no longer climb the stairs to his and my mother's bedroom, and their bed will need to be brought downstairs. My parents, I'm sorry to say, have made a career out of living in the moment, putting off anything that can be put off another day. Except my mother finally understands that stair-climbing time is about to run out.

They call this room the "salesroom," the vision having been that customers to their junkyard might enter this room and not the living quarters of the house. The room was never finished, however, it's floor still a concrete slab somewhere below the flotsam. The salesroom and its attached half-bath long ago became packrooms, which my mother and I are unpacking.

The odor of mold is so intense it might actually overpower one of us. It seems deadly.

"This mold can't be good for you," I say.

"We need to get the door open," Mama says. "And a fan going." Never tall, my mother has lost some height and earned some wrinkles. Her soft hair is light gray now, tucked up with pins. She stumbles a little as she turns to retrieve an electric fan. She wants to help me with an extra door, a random white door leaning against the actual brown door on hinges, but it's blocked by a whimsicality of stuff and is too heavy for her anyway. I wade into a thicket of walking canes, maybe five of them, handing them two at a time to my mother, who plants them in a cane forest by the large fireplace. I move a wooden clock with a chain hanging out the bottom, careful not to hit a dusty, foot-tall glass case, its walnut frames loose, displaying a wadded piece of paper. I stand holding the flimsy case, surveying for a horizontal clearing to stash it.

Finally the door is open, sunlight filtering in, the September day heating up southern Georgia. The fan is humming.

My parents are collectors. My sister says to get real and call it by its real name, hoarding. I close my mouth tight.

There is an entire banana box of chargers — for cell phones, cordless phones, toys, tools. If somebody needed a charger, this would be the place to look. If someone needed a table lamp, this would be the place too. I had no idea my mother loves lamps this outlandishly; it's almost a fetish. One lamp's glass base is filled with seashells. One is large, ugly, ceramic. Somebody has ruined a crockery jug making another. Some have torn shades, some have intact ones, some have none. Five or six shades are stacked inside each other on an antique bed.

"We've got to figure out what to do with all these lamps," I say to my mom. She's standing just outside the door ripping rotten cloth off a lampshade. She intends to save the metal skeleton of it. Her sensibilities would never allow her to pitch the entire ruined shade in the trash, since she knows very well from life on a junkyard that the metal can be recycled and that recycling metal is lucrative. "Got any ideas?"

"I don't, honey. We'll have to pack them up somewhere."

"Do you know anyone who sells lamps? You could give them all away. Somebody might be able to make money on them." Even as I say this, I know in my heart that's not going to happen. My folks don't give anything away. They are keepers.

"We can put them under the old car shelter out at the farm," Mama says.

"They'll just ruin outside," I say. "But we'd have a lot more room to work if we got them out of here." There are at least twenty table lamps. I begin to line them on a little patch of floor at one end of the room, between two large rolled brown carpets, half-painted canvases, lone shoes, a wicker laundry basket filled with somebody else's clothes, all manner of etcetera. This clears a small island of floor space, enough to start packing up the boxes I've brought.

I begin packing whatever is within reach — an alarm clock, a basket of silk flowers, two candle holders, a ceramic owl, a wooden owl, a brass vase, a conch shell. Many of the things I find are ruined —

dusty, molded, crushed by weight, decomposing — because, I realize, when a person is a collector, when his or her mission is collecting, maintenance is not an objective. My mother is hesitant to throw out anything, second-guessing herself to think of a use, but I take a break from the stuff and go roll their curbside trash receptacle to the edge of the unfinished porch. It gets the rotten pantyhose, the broken glass, the shoes warped and split beyond use.

I keep packing, stacking boxes in white space I've liberated along a wall. "Two more canes," I say to my mom.

I find a handful of pencils and pens. "Let's put office supplies in one box."

"Here's a bag of pens." My mom hands them to me. I would bet good money that 90 percent of them don't write. My mother is working hard. She knows that it's time.

We don't get far before we find termite damage. My parents knew we'd get to it. They discovered termites a decade earlier and treated the house foundation and the footing.

"So that stopped them?"

"That stopped them," Mama says. "We just never got the mess cleaned up."

My brother shows up at the door. In his early fifties, he's tall, with his still-black hair trimmed short in the front, growing past his shoulders in the back. He follows me outside to the trash receptacle into which I'm tossing cardboard that termites have turned to powder.

"Do you think the termites are gone?" he asks.

"I don't know."

"I wonder. If not, the whole house could fall down around them."

"God, let's hope not."

"Did you see how the termites ate parts of that wooden desk?"

"No. But I see what they've done to the books."

My father has bought a couple thousand copies of my second book, in hardcover, sold to him when the paperback was released, and they are stacked at the back of the salesroom in boxes. Daddy can't resist a deal and he doesn't want to see my name on the remaindered shelf. But termites have been enjoying the books a lot more than readers have. A few hundred of the books, it seems. Maybe more.

Eastern subterranean termites, the most common termite in the southern United States, work like this. They eat dead plant material, not only the structural timbers of buildings but also—once they break and enter a place—paper, cloth, carpets, photographs on the wall. They construct little tunnels through which they travel, called shelter tubes, which keep them hydrated and out of the sun and protected from predators, including humans. These tubes are made of plant matter, saliva, and soil. The termites tunnel up from the earth, where they live, through soft wood or some other biodegradable substrate, and they tiptoe through your property, chewing on what they can digest, which is cellulose, meaning wood or paper or cardboard, skulking inside their mud tubes.

They are detritivores, the scientific papers say. They live on detritus. They turn it into feces, golden pellets of droppings.

That means they are making dirt.

My own home, Red Earth Farm, is on a dirt road about thirty miles from my childhood home, where my parents live. In our garden shed is a jarful of dirt. When my husband and I give tours to people who want to see a working farm—not an industrial monoculture row crop farm but a farm circa 1900 with its milch cow and its hogs, its guineas and its Jersey bull, I lead them to the doorway of the shed.

"This is why we call the place Red Earth," I say. Because of the clay at the bottom of the dried-up well, silky and moist but tough, resisting, a substrate of everything we do. Because of the native people who fought for this land until the creek ran red. Because of the slaves and the blood they lost. Because of the blood that ran from the women, and the men, and the animals whose lives were given to the people.

It just looks like dirt that happens to be red. Like a jar of crumbly loam. Like the frass of the termites; however, it is a sacred thing, constructed of the detritus of so many living things that inhabited this land. It is made of life and it is alive, both.

When the visitors are gone, I walk the dirt road alongside the farm the mile to the mailbox. Along it I find shards of flint left by the Creek inhabitants. On almost every walk I find something. We don't have flint geologically in southern Georgia. We have ferrous oxide, little red rocks called "rain rocks" because rain washes them out. The lighter flint (lighter in shade and in weight) is easy to see among the scatterings of blood-red pebbles. I bring the shards home in my pockets, gathering up fragments of points and blades because this stuff belonged to people no longer here to pack it up. I place the artifacts in a flat basket woven of longleaf pine needles. In the basket is a button that says, "Go Make History."

I try not to get depressed about all the stuff in my parents' home. I have a hate/hate relationship with it. My parents, both born poor, came from a time in which stuff was hard to come by; now we live in a world lean in the things that really matter and that really fulfill, like civic engagement and family time and neighborliness, but fat with stuff—until all of us as consumers are morbidly obese. Stuff is constantly being thrown at us. As an author, I get more than my share, besieged with gifts everywhere I travel—cloth bags and T-shirts and pens and books. Book after book after book. We humans can't quit making stuff, using more and more of nature to make more and more material crap.

Whereas the children of the Depression, like my folks, had to work hard to accumulate possessions, we have to work hard not to accumulate. I have to send a little note ahead to say, *Please, no gifts* to the folks who have invited me to do talks. I have to constantly walk through the rooms of my house with an eye turned toward purging.

So many human interactions are based on things, swapping our bounty back and forth, eBay, Craigslist, the classifieds, flea markets, consignments; and all shopping in general, of course, is simply human interaction at its most primal. But we have enough gewgaws and whatnots and doodads. No more. I have to continually seek different kinds of interactions, a different focus.

One Saturday during the clearing-out time at my parents' home, our neighbor, who is eighty-two and also paring down, gives us all his worms. He's been keeping the worms that make dirt for him for fifteen years and he has lost interest. Besides, the wooden tubs he built for them are rotting and he'll have to construct more if he doesn't divest. Better to give them to the younger farmers down the road, my husband and me, whose enterprise is to make dirt, to build healthy soil in which healthy food will grow. This means soil full of microorganisms, full of minerals, full of mycorrhizae.

"They'll eat just about anything," Bill tells us. "Except citrus. They don't like citrus. You got a paper shredder? They'll eat paper."

"What about eggshells?" my husband asks.

"Eggshells," Bill nods. "I gave them some peanut shells once. It took them a long time to get rid of those. But they did it." He tells us to bury the food to make it more available to the worms, since, averse to light, they hesitate rising to the surface of their bin to haul down food.

I run my hands through the awesome loam the worms have deposited. I hit a half-rotten pear and a pod of oversized okra. The okra holds its shape, but it is full of worms, not plant fiber, and they are busily turning the okra into something better than it was, some endgame we're all heading toward, some substance that is the essence of all life. I bury the worm-pod and gather up a handful of gummy earthworms wrapped around clumps of rich black soil, rich as my parents' packroom, rich as the bottom of a termite colony. Praise the transformers. We all become something better than what we were.

I've decided three things: one, I will accept anything that my mother decides to give me on the days that we clear out her salesroom. I will take any gift home to the farm and do with it what needs to be done.

Two, I will do this job without judgment. My brother and I have a running conversation, the basic summary of which is, "Oh my God, what are we going to do?" That conversation begins with my parents' long-standing date with the yard sales on Saturday mornings; or with the clearing out of a section of their yard that immediately begins to be refilled; or with the observation that their pack spaces are not efficiently filled, but are organized jammed at the doors, until the doorways become blockages, as if one or the other of my dear parents goes to that doorway and sticks something new through it, hoping only to get the door closed before the avalanche. My brother and I want our parents to change their ways. Magically. Which we realize, of course, is not going to happen.

Doing this cleanup job without judgment means just doing what needs to be done, without getting bogged down in the past or the future.

The third decision is really an epiphany. On a second or third visit, slowly emptying the jam-packed room, my father and I have a candid conversation about the termite-drilled books. "Let me have the ruined ones," I say. "I'll take them home and burn them." (To make ash, also used to make soil.)

"It's almost impossible to burn a book," my father says.

"Not enough oxidation?"

"You'd probably do better to bury them than burn them," he says.

In the end, however, he won't part with the books. The undamaged ones will stay in the salesroom, still in boxes along the back wall. When my parents move their bedroom downstairs, they will eye the books daily. The rest, the damaged books, will go into an old trailer out at the farm that my folks also use as a pack house.

My father's knees will not allow him to transport the books. I will do it for him, packing up termite-tunneled, chewed, powdery, unreadable books as if they were bound for the Smithsonian, stacking boxes ever so carefully at the back of the room, high as the ceiling, working my way toward the door.

You know what I'm thinking. I'm thinking that termites can take a book back down from words and ideas and pages to nothing. No, not nothing. Back down to dirt. Where it all started.

That richness leads me to my third decision: I am not going to worry about what will happen to all the stuff my parents have collected, or that I have collected for that matter, because it's all going back to the earth. All of it. The termites, the worms, the millions of bacteria in a spoonful of soil—they will happily do the work of deconstructing what we have constructed, and valiantly. Working as emissaries of the dirt, they will reclaim most of what humans have fabricated, at least the plant-based material, hauling it earthward in the tiny wheelbarrows of their mouths, one wee load at a time. Bacteria and rust and time will get the rest.

The little creatures will take down the books, they will take down the book jackets, they will take down the cardboard boxes. They will take down the shades, they will take down the canes, they will take down the walls and the roof.

All the stuff goes back.

Praise to the little beings who are the transformers. Praise to the transporters. And praise to the ground that accepts and banks it all, that keeps rising out of lava and bedrock into layers of humus, unsurpassed transformation, the best savings account humanity could have, the one real thing that stands between us and scarcity, poverty, hunger. Praise to the dirt, which contains all that we have and all that we are, our true home, our destiny.

GLOSSES ON DIRT

❧

ERICA OLSEN

*A*eolian. It has something to do with wind. The aeolian harp, its strings plucked by wind; the Aeolian Islands, off the coast of Sicily, in a part of the world where mythology governs the naming of places. Usually, I'm confident about the words I use, and I bend their meaning this way, or that way—in any case, to my will. But when it comes to words archaeologists use, I'm not always certain of their meaning. *Aeolian* is one of those words. It sends me on a little expedition to the dictionary. A curator at an archaeology museum once told me that when cataloging artifacts, it is not acceptable to make aesthetic judgments. In a collections management database, description means description. Describe the projectile point, its type—for instance, corner-notched—and the stone from which it was knapped, but if it's pretty, don't call it pretty. Creative writing and archaeology cannot be done in a mutually satisfactory way. I worked for six months as the lab manager for a contract archaeology firm that was excavating ancestral Pueblo sites at the White Mesa uranium mill south of Blanding, Utah. One week, when the field crew was short-handed, I helped screen for artifacts in a field of aeolian—which is to say, *wind-blown*—soil. Aeolus was the Greek god of the winds. You see my problem.

Artifacts. At my job, I spent hours washing artifacts. Mostly I washed potsherds, fragments of the broken bowls, seed jars, and other vessels of the Pueblo Indians who lived in southeast Utah a thousand or more years ago. In the lab, which was not a real lab but rather a rented garage with no sink, I'd fill a bucket with water from the garden hose and scrub the sherds with a toothbrush. I washed stone scrapers, hammer stones, manos, and the fragments of shaped, smoothed sandstone that the archaeologists called bulk indeterminate ground stone. I washed a quartz crystal. I dry-brushed the artifacts I couldn't immerse: awls made from animal bone, fragments of turkey eggshell, and rosy sherds whose color, applied after firing, was known as fugitive red. Washing artifacts was simple, physical work, and I liked it. Each artifact told a story of the past, but it didn't tell it to me. We occupied adjacent spaces, the artifacts and I, in companionable silence, like seatmates on a train in a country where I did not speak the language. Where I work now, at an archaeological research and education center, a visitor asks if washing removes valuable residue from the artifacts. Yes, I reply, but not everything warrants that level of analysis.

Flotation. The archaeologists went off early each morning to the excavation site on White Mesa. At the end of the day they brought back brown paper bags of artifacts and samples, labeled with the precise location where they were found. The collections included numerous one-liter bags of dirt, scooped from the twenty archaeological sites they were excavating—an area that soon would be destroyed in the construction of a new "cell," a holding pit for uranium mill waste. That construction project was the reason for the archaeology, which bore the straightforward name of the White Mesa Mill Cell 4B Data Recovery Project. In the lab, the soil samples were processed by a college student, a theater major, the daughter of the archaeologists who owned the firm. Kelly did this work in the driveway of the garage, using a garden hose, buckets of water, kitchen strainers, and other homely implements and wearing her bathing suit because thirty years earlier her mother had processed floats similarly attired. When her hair was up, you could read the tattoo in delicate script on the back of her neck: *So much beauty in dirt.* Successive pour-offs of black sludge resulted in the separation of what archaeologists call the heavy fraction and the light fraction. What came out of the muddy water was left to drip-dry in pouches of diaphanous white mesh, like little ghosts suspended from a clothesline that stretched across the back of the garage. This material would later be examined for microartifacts—small particles of animal bone, plant material, delicate ornaments such as shell beads—that would have passed through the ¼-inch screens in the field. "It's not what you find, it's what you find out," archaeologist David Hurst Thomas has said. If we look at dirt closely enough, we are no longer looking at dirt. We are looking at human lives.

Loess. Wind blows dust, and dust settles into loess, a word of German origin that means a silty sediment. Aeolian loess soils were deposited after the last ice age across much of southeast Utah and southwest Colorado. In a region known for its canyons, loess is rock's unglamorous cousin. It can't compare with the fine bones of sandstone, the arches and the natural bridges. Loess sat unappreciated—hunter-gatherers didn't find a diversity of plants suitable for foraging there —until, the archaeologists say, people began farming corn and beans and realized the deep soils held moisture in the high desert. The first time I walked through Utah's aeolian loess I was in my twenties, and the place gave off something like pheromones, that jolt you get in the presence of an attractive stranger when you admit, to your joy or shame: *I want this.* Now in my forties I analyze the attraction "objectively." There seemed to be space for me here, in this place. But I did not figure in the degree to which I was myself aeolian.

It was hard to love the loess at the mill. The wind blew steadily from the southwest, stirring up dirt devils at the excavation and the adjacent construction site where another kind of crew was busy obliterating what the archaeologists were done excavating. The wind drove grit the texture of cornmeal and the color of cinnamon into our ears and nostrils. Inside the portable toilet, a fine powder of loess covered every surface, including the After the Breakup Wheel someone had left there for the amusement of the field crew. You could spin the dial, then choose an action to take: revenge, wallowing, and so forth. Any fingerprints left on the wheel would soon be covered up by another layer of dust. In their work clothes, you couldn't tell the girls from the guys. But on their field specimen bags, the girls were the ones who wrote "Pretty!" when something was pretty. The archaeologists indulged in black humor about the contaminated soil at the mill and about the toxic pond where yellow-stained water arced from the aerating fountains and a fake bald eagle failed to keep migratory birds away. Before the van in which we carpooled drove off the mill property, it stopped at a washing station. The wrong kind of dirt was not to leave the property. My first day on the site, I expected an enclosed, sterile facility. Instead, we rolled up the windows, and two Navajo guys in ponchos sprayed us down with hoses.

Midden. A lovely word for a trash mound, with comforting echoes, *middle* and *den.* I learned to tell a midden by the presence of gray soil — gray from ash swept from old hearths. I can recognize that ashy soil myself, now, when I'm out hiking. Other than middens, I'm illiterate when it comes to reading dirt, even though I've watched archaeologists read it many times. They see things in the earth of early spring when it is wet with snowmelt that would be impossible to see in the same earth baked dry by summer heat. One archaeologist may spot something that another archaeologist may miss. One may, I suppose, call the soil in a unit loamy silt, while another calls it silty loam. But no one can fail to identify a midden.

Munsell color system. Used by archaeologists, soil scientists, and others who are intimate with dirt. The Munsell system uses letters and numbers to represent color in terms of hue, value, and chroma perceptible to the human eye. The warm-spice-colored earth I fell in love with: 2.5YR 5/3. The bare soil that strikes my eye as drab when I am sad: 2.5YR 5/3.

Pithouse. The artist Rachel Whiteread casts negative spaces. *Shallow Breath* is the space beneath a bed. *House* was a concrete cast, in 1993, of an entire Victorian house in London before its scheduled demolition in 1994. For me, archaeology was like that: a revelation of spaces. At the site on White Mesa, the places where soil changed color guided the archaeologists in their excavation of postholes, storage bins, hearths, and other features that combined to delineate the shape of seventh-century Basketmaker pithouses. At the excavation, the sight of a pithouse, or more accurately the presence of one, with its clearly defined area for cooking and its equipment for grinding corn, affected me greatly. Here was a home, snug and earthbound, with a place for everything and everything in its place, *in situ* as the archaeologists say. I, on the other hand, kept my essentials in my car, *in auto*, so to speak, and everything else in two storage units, one just up the road in Blanding, Utah, the other in Bellingham, Washington. But if the pithouse I saw had been found with its grinding stones left carelessly askew, that would also have affected me greatly, although in a different way.

Pottery. The week I spent screening on White Mesa, I worked as hard physically as I'd ever worked in my office-bound life. I was living in a borrowed place, an old pottery studio that was equipped with a toilet and a utility sink but no kitchen or shower. In the morning, before joining the van pool up to the site, I'd go to the K&C store and buy a pack of Hostess Donettes. I'd eat half for breakfast and save the other half for lunch. I'd swallow my antidepressant. During that time, I was unhappy. Now, I am not unhappy. My feelings are not what matter, about the work I did then, or the artifacts. Description means description. Potsherds made me think about containers — about the ones the archaeologists used. They moved dirt with scoops they'd made out of cut-up one-gallon plastic jugs. Metal clipboard boxes kept dirt off their forms in the field. It all began with baked clay: dirt made into containers for things you want to keep dirt out of. We still live on the path the Neolithic revolution put us on, the transition from hunting and gathering to agriculture and a settled, sedentary way of life. We store food. We save for the future. By *we*, I don't mean *me*. By Neolithic standards, I was inadequate.

Samples. In addition to soil samples, the archaeologists took pollen samples, which they folded into packets of aluminum foil, and dendrochronology samples, sections of tree limbs that they wrapped around and around in white cotton string. Working in the archaeologists' office that winter was like working in the nineteenth century. First thing in the morning Tamara would sweep while I got the fire going, or I would sweep while Tamara got the fire going. Sometimes I brought a potato to work and baked it, wrapped in foil, in the wood stove. I made sure not to get my foil packet mixed up with the other foil packets. Mine was for lunch, theirs were for science.

Screening. My screen looked like a rough-hewn wood-framed window screen slung waist high from chains attached to a collapsible metal tripod, balanced on uneven ground, splay-legged over a mound of screened dirt. I'd heave a five-gallon bucket of dirt onto the screen, take hold of the frame, and commence a vigorous shaking, back and forth, then side to side, turning my head to the side as dirt poured through the screen and rose airborne in clouds of choking dust. I liked the screening, just as I liked the washing. Sometimes dirt fell away from artifacts during the first shake. It was easy then to pick out the sherds and lithics and drop them into my paper field specimen bag. What was left in the screen, the rock-hard clods of dirt, had to go through the wire mesh as well. That meant anything from lightly tapping at the clods to whacking them with a trowel or, more often, a sawn-off section of broom handle. You could get a rhythm going. This is another thing I learned from working with archaeologists: there is the dirt you treat with tenderness, and there is the dirt you can hit.

Stratigraphy. A stratigraphy diagram shows levels of deposition and levels of occupation, layer by layer, down to ancient periods of human habitation. Stratigraphy is reassuring. Things that were on the surface are no longer on the surface. Things that used to trouble me now trouble me hardly at all.

Temper. In the lab, I washed potsherds to remove the dirt, and when the sherds were dry I counted them, weighed them, and put them into archival-quality polyethylene bags. Eventually, it would be someone else's job to open the bags and analyze the sherds. If you are a pottery analyst, one of the things you look at is temper: the materials mixed in with the clay, which show at the broken edges of a sherd. Temper can be gritty, like sand or crushed rock. Temper can be crushed sherds. Once, when I was washing artifacts, a grayware sherd, badly tempered or badly fired or both, softened into mud in my hands.

My own temperament includes, by now, a good amount of southeast Utah dirt, which can be described—though no one, not even myself, can possibly succeed in analyzing it.

SOIL VERSUS DIRT

A Reverie on Getting Down to Earth

*

KAYANN SHORT

"*N*ice to be able to till, huh?" I wave at John as he drives the old tractor through the bean patch he's turning. He nods and smiles, happy to be on the John Deere again. After record rainfall in September, the soil John just tilled for next spring's crops is perfect —as silky as fine powder to the touch, but with enough humus and moisture to stick together. We are an organic farm that amends our soil naturally with compost and "green manure" cover crops of rye, field peas, oats, clover, and buckwheat for turning into the vegetable beds between plantings. Our soil is our livelihood, one we should never take for granted. Or, as my students would say, "for granite." Which in the case of soil is appropriate, since soil started as rock millions of years ago. If it were granite, we couldn't grow here.

When I first met John many years ago, he invited me to his farm, Stonebridge—ten lush acres of land that had been farmed organically for a hundred years along Colorado's Front Range. On one of our first farm dates, we worked together in the greenhouse, making blocks for new spring seeds. I remarked how good it felt to get my hands in some dirt. John stopped and looked up from the flat he was seeding. "Call it soil."

"Okay," I agreed, but in my head, I rolled my eyes. John loves math, and I love literature, so we think about words differently. He likes them to have a solid footing; I like a little more sway. I knew from an agricultural perspective he had a point, but I was irked to find it aimed toward me. My grandparents were farmers and I'd had many gardens of my own. I'd been on the earthy side of things my entire life. Whether I'd had my hands in soil or in dirt, what mattered most to me was what would grow there, not what it was called.

I've thought a lot about that conversation over the years John and

I have farmed together at Stonebridge, and I still haven't teased it all out. My annoyance came from being corrected, but also from sensing a deeper divide between the two terms, one I've always straddled as a writer, farmer, and activist. To me, "soil" can sound kind of snooty, more like something sterile from a bag than the rich stuff in our own backyards. But "dirt" might not say enough about the way organisms help us grow. What is the difference between soil and dirt—and does it really matter?

As John's remark suggests, "soil" is often a term for something living, a material substance full of life forms that contribute to the land's ability to produce food or raise livestock. As the term used by agronomists or horticulturalists to mean the upper mantle of millennia-old decomposed rock, or "topsoil," found on the earth's surface, "soil" evokes a scientific sense of organic matter that can be quantified or is productive in a measurable way. I find that for most people "soil" is a positive term because it's associated with gardens and crops and food. We do use "soiled" to mean ruined, as in reputations, or needing washing, as in laundry and diapers, but generally "soil" refers favorably to something beneficial, rich, and fertile.

"Dirt," on the other hand, runs a whole gamut of values. If something is "dirty," it could be untidy, spoiled, or polluted, or "dirty" could mean salacious, like X-rated magazines or movies. Just as "weeds" are plants growing where they're not wanted, "dirt" gets vacuumed, shoveled, and swept away in an effort to make things clean. Yet paradoxically, there's something wholesome about the ubiquity of dirt as it crosses social boundaries and connects us to the ground beneath our feet.

In *A Bushel's Worth: An Ecobiography*, I pondered the difference between dirt and soil in a discussion of farms as "cultivated space" where humans and nature form a fertile alliance: "[I]n farming, the word 'dirt' usually refers to something devoid of life. While soil is alive with organisms in a food web that nourishes us through helping plants grow, dirt is considered something to be removed, scraped out by bulldozers in new housing developments or cleaned out from under your fingernails. If you're a farmer, though, you aspire to dirty fingernails." Simply put, in farming terms, soil has something growing in it and dirt doesn't.

My grandparents and great-grandparents farmed in North Dakota during the Dust Bowl period they called the "Dirty Thirties," no doubt a term that reinforced our country's collective disdain for dirt. To mitigate the dust storms born of drought and "plow-out" farming practices that stripped the land of native grasses holding moisture and preventing soil loss, the U.S. government implemented erosion programs through the Soil Conservation Service, things like strip-cropping, contour plowing, crop rotation, and the planting of wind breaks. Both my grandparents' farms had large groves of trees near their houses, long rows of Russian olive, Chinese and American elms, and blue spruce that broke the prairie winds.

During childhood vacations to the farms, I loved wandering the windbreaks where the ground was soft from my grandfathers' tilling and the birds sang in the cool shade. I didn't know these canopied rows had been planted as a conservation measure. I just thought my grandparents wanted some trees to break up the monotony of wheat fields and grass.

I don't remember my grandfathers using the word "soil" for the pastures or fields on which they farmed, but I do remember my grandmothers lamenting the dirt that got tracked in on everyone's shoes and had needed daily sweeping from the floors. My mother corroborates my memory of dirt as my grandmothers' burden: "Dirt was the topsoil that blew around when it was windy. No one liked dirt, especially on the windowsills." But unlike me, my mother remembers my grandparents' worried talk of "dry soil" or "wet soil," depending on the moisture that fell on the fields and determined when crops could be planted. Farmers couldn't plant in the spring until the soil dried out from the deep North Dakota snows, but they also needed rain to germinate the seeds and water in the crops in non-irrigated, dryland fields.

My cousin remembers our Grandma Smith using the word "soil" for the pots of African violets she grew near the picture window and "dirt" when she handed the grandkids shovels to play in the rocky yard between the farmhouse and barns. Dirt was also where flowers got planted on the highway-facing side of the farmhouse — hollyhocks, sweet peas, and poppies to brighten the day.

My dad, who grew up driving tractors and plowing fields, insists dirt and soil are the same. What he means is that farmers didn't need to differentiate between the two words because everyone knew what they meant: hard work on the land raising crops to put money in the bank for next season's seeds.

As a farmer, I know I should come down on the side of "soil," and I do use that word frequently for its connotation of biodiversity and organic matter. But I have to say I'm sympathetic with the idea that "dirt" and "soil" mean the same thing, or at least can be used interchangeably. I don't want to bolster a division between experts and the rest of us, nor do I think one word should stand for good and the other bad. Besides, with its double Ds, "digging in dirt" sounds more fun than "digging in soil." As long as we're digging, something must be growing, so what we call what it's growing *in* shouldn't really matter.

As a writer, I know the play between "soil" and "dirt" is useful, as in this *National Geographic* title, "Are We Treating Our Soil Like Dirt?" But I wonder whether our abandonment of "dirt" to something lowly, dead, or detrimental points to the disregard for nature that drives our larger environmental problems. If we reclaim dirt as living, maybe we'll quit pumping it full of toxins and waste. Maybe we'll conserve it through better farming practices that reduce cultivation on marginal land, despite high prices for crops. If soil is something we revere and protect, shouldn't we be treating our dirt like soil?

On the other hand, the double entendre of "treating" — "applying additives," as well as "behaving toward" — is operative in industrialized agriculture today with its dependence on oil-derived and expensive inputs to boost deficient farming practices. It's soil that's doused with chemical fertilizers, herbicides, and pesticides, soil that's chemically stripped of organic nutrients and erosion-preventing vegetative cover. If *this* is how we treat soil, maybe it would be better off as dirt.

Thoreau uses the term "soil" when writing of the moisture and fertility of his bean fields, but he also uses "earth" to mean the loamy material in which plants grow. Any farmer or gardener can relate to

Thoreau "making the earth say beans instead of grass." When he refers to his bean field as "this portion of the earth's surface," we understand how minuscule a harvest his labors will yield within the vaster universe in which his field resides.

Maybe "earth" is the word to reclaim for that substance on which our lives depend. As a synonym for "planet," "ground," "dirt," and "soil," in all their various forms, "earth" seems to do it all. Isn't "down to earth" a compliment, a term for people we admire who don't have fancy notions or, in elemental contrast, put on airs? Down-to-earth people are *grounded* and can be trusted to do the things they say they'll do. They're dependable, like soil, like dirt, like the things we take for granted—or granite.

Several years ago, the seed co-op to which we belong included this anonymous quote in its yearly catalog: "Humans—despite their artistic pretensions, their sophistication, and their many accomplishments—owe their existence to a six-inch layer of topsoil and the fact that it rains." Earth is the evolutionary origin and outcome of ancient cataclysm between rock, wind, and water, geology and cosmology forming the place on which all of us make our home. Hard to imagine how such a thin crust could nourish seven billion of us, not to mention the multitudes of other organisms living by our sides.

Soon, the winter days will lengthen, the sun will warm, and John and I will return to the greenhouse to start our spring crops. In the screen-covered box we call the soil maker, we'll mix the ingredients for the organic material in which we plant our seeds: one half-part peat that we purchase, one half-part willow root harvested from our trees, one part sand and silt washed downstream in mountain runoff to our waterways, and one part compost from the loamy pile we tend behind the barn.

This year, I'm going to call our mixture "earth" in honor of the ancient forces behind each Stonebridge handful of mingled water, land, and air. A microcosm of our farm's living matter and, beyond that, the world's, this earth John and I make offers the necessary ingredients to start seeds for a new season. Once they sprout, we'll transplant the seedlings into more spacious earth where they will grow.

Our farm is just ten acres among thirty-six trillion others, but we grow a lot of food in this cultivated space. We'll keep adding organic matter, preserving trees and grasses, and tilling responsibly to protect Stonebridge's future. I've come to use the word "soil" more often than I used to, but "dirt" is still in my farm vocabulary, especially when I'm digging out the thick roots of perennial grass that crisscross our land. A little of my versatility has rubbed off on John, too. Working with the crew the other day in the fields, he said "dirt" and we all laughed. A farm is many things — topography, crops, wildlife, water — but underlying all of it is the rich, lively earth that sustains our work. Whether John calls it *soil* or I call it *dirt*, in scoop, bucket, field, terrain, and tilth, we'll keep farming with our hands in it together.

DIGGING IN

*

ELIAS AMIDON

*I*n the 1960s the poet Gary Snyder advised, "Find your place on the planet and dig in." I took his advice seriously, digging out roots to clear gardens, digging holes to build fences, digging trenches for foundations, drainage pipes, and septic systems. The view from the handle of my shovel helped me love each place I made home, and the sound of my neighbors' shovels working next to me did the same.

By learning to dig in — literally and figuratively — I became a neighbor: living under the same weather, walking on the same ground, sharing the patience needed to live in a place and care for it. But the borders of my home place have never remained secure. The fences I've built get climbed over, the gates get left open, and the world finds its way in. Faces stare out from the newspaper page and TV news — the taut faces of hungry people half a world away. Children on street corners in desert towns watching soldiers pass by. Faces and voices and stories unknown to me. *Who are these people? Whose neighbors are they? What is important to them? Why is their land under siege?*

These questions have made me put down my shovel more than once and go on the road, more like a pilgrim or wayfarer than a tourist. Some outer, credible purpose usually motivates my journey, but inside I am waiting for that moment when the ground shifts — when humor or sadness or a shared recognition makes the distance between "my" world and "the" world beyond my borders vanish, at least for a moment.

Once, in Bethlehem, I was invited for a midday meal at the home of a Palestinian family. Their flat, perched on top of a three-story hillside building, had been in the direct line of fire between Israeli forces and Palestinian fighters holed up in another building on the hilltop. Israeli bullets lodged in their walls.

My host told me their old refrigerator had been punctured by machine gun bullets, causing it to defrost and the meat in the freezer to thaw, leaking a red stain across the floor. Their three-year-old son had pointed at it and said, "Look Daddy, they killed the fridge!"

We all laughed. Then we were silent, our eyes turned to the stain. I glimpsed the world with a three-year-old's eyes. Yes, I was just a stranger passing through, but I became their neighbor too in that moment. Digging in to a place as they had, and traveling through as a wayfarer as I was doing, are about the same desire: *to bring the world close.* But they do this in markedly different ways.

The indigenous villagers of the Pageiyaw tribe of northern Thailand—whom I visited and worked with for a decade—mark the span of their lives by literally digging in to the soil. They bury their dead among the trees they call "the ancestor forest," and in an adjacent area known as "the umbilical cord forest" they place the placenta of newborns precariously on a tree branch. After some days, when the placenta falls to the ground and dissolves into the soil, it means the infant will survive and will henceforth belong to that land all its life.

In contrast, the wayfarer doesn't bring the world close by anchoring to a home place, but by listening to its stories as he or she moves through, and then telling those stories to other people in other places. This listening and telling was also the work of the ancient bards. To help the ones who have dug in avoid becoming suspicious of people not of their place. It is a brand of peace making—at least this is my belief and my hope. The wayfarer is no tourist: she doesn't stand back but enters in, asks caring questions, listens, and shares. She weaves the world together, bringing all lands close with her caring attention.

For a number of years I worked on a project called the Abraham Path—a meandering dirt path that follows the proverbial footsteps of Abraham from the ruins of Harran in southern Turkey, where he was "sent forth," through Syria, Jordan, Palestine, Israel, to the town of Hebron, where Abraham and Sara have their resting place. Hospitality and kindness to strangers are hallmarks of walking this sacred ground. The path is now well established in many parts of the Middle East, and for a while it looked like the Syrian government would allow pilgrims to pass through. But then the peace-making power of such a path worried the Syrian government. They realized this kind of cross-border path would allow a free flow of wayfarers across the land; it would bring people who walked on the ground like pilgrims do, vulnerable and open. Such pilgrims would meet each other face to face, and tell each other their stories. This seemed too risky to the Syrian power holders. After nine trips to Syria, I was blacklisted and not allowed to enter again.

Over the past twenty-five years I've filled my passports with hundreds of stamps marking border crossings. Crossing through the checkpoints between Israel and the West Bank are not marked in those pages, but those crossings are some of the most vivid in my memory. As a wayfaring American, I could pass where people who were dug in on one side of the line could not easily cross to the other. Though they lived only a few miles apart, they never met. The bleeding refrigerator in Bethlehem had belonged to Palestinians. I also visited, on that same trip, an isolated hilltop enclave of Orthodox Jewish settlers established in the middle of the West Bank. I wanted to speak with the old rabbi who founded the settlement, a well-known teacher of mystical Jewish texts. By the time I arrived a cold wind blew intermittent rain over the dark hills. The place was dug in like a fortress, isolated from the land around it. My Palestinian driver left me at the gate, saying he did not feel he should be there. He told me to call a taxi when I was ready to leave.

The several young families who had come for teachings from the rabbi that evening were trying to quiet their fussing children, who were clearly ready for bed. The rabbi—with long white hair and beard, bushy black eyebrows arching over kind eyes—gave them blessings, a scene of intimate tenderness. One by one the young families left. The weary rabbi seemed genuinely disturbed that my Palestinian driver had not felt comfortable visiting, even though he was his neighbor. And then he told me of his deeper sadness. The Israeli prime minister had just slated his settlement for evacuation. Twenty-seven years of digging in would simply end. I thought of Gary Snyder and how all boundaries bend to time.

The rabbi walked me back to the guardhouse to meet my taxi. He asked the young soldier if I could wait in his little tin guardhouse out of the cold. The wind moaned through the cracks in the squalid hut. Hebrew graffiti were scrawled on the walls, and one in English read, *"What am I doing here? What am I doing here?"* The Israeli soldier stood there silently, with his machine gun slung over his shoulder, looking down the hill for anything that might be approaching. I felt the neighbors there, so close yet un-meeting. I felt the little Orthodox children drifting off to sleep nearby, and the child of my Palestinian friend dreaming of the murdered refrigerator. The graffiti echoed, *"What am I doing here? Why am I not in my own peaceful home far from here?"*

But I already knew the answer. I was simply supposed to be there, feeling exactly this bittersweet recognition that I repeat now: the ground we dig into and walk upon is sacred. It is sacred because it makes us neighbors to each other, whether we like it or not. Tell this story.

2

KID
STUFF

major in mud pies

THEY TEACH ANYTHING IN UNIVERSITIES TODAY.
YOU CAN MAJOR IN MUD PIES.

ORSON WELLES

DIRT PRINCESS

*

JULENE BAIR

$\mathcal{S}$ometimes I banged around the farmyard on the heels of my
cowboy boots, seeing if I could make it the whole way between
the shop and the house without lowering my toes. Sometimes I
went barefoot. My toes gripped the hard-packed clay, spread over its
knobby surface, and, brushing back and forth, managed to raise a
powder. One summer I wore sandals, little pink-soled princess things
with glitter embedded in the clear plastic uppers. Wearing them, I
felt the stir of silk and lace about me, and I danced in the dust of a
fairy's wand. All I was really clad in were jean shorts and one of the
sleeveless blouses my mother made for everyday—my tawny, blond-
furred skin marred by scabs and scrapes and my spindly legs and
arms streaked in watermelon juice and horse snot.

I've spent forty years getting the connection straight. I didn't come
from Adam's rib but from the clay, just like him. I didn't need an inter-
preter to talk to God. My body rubbed against the dirt of that farmyard
when I began to crawl. I felt its hard humps beneath my hands and
knees, its stubborn solidity in my belly when I fell flat, or sprawled for
pleasure. I ate bits of it and tested other things that rested or crawled
upon it—fire ants, rocks, kernels of grain fallen off the trucks. I raked
it and made roads and fields for cars and tractors. I wrote my name in
it, caught toads in the mud ruts when it rained, played roll-over dead
with the dog on it, mewed and followed mother cats on all fours over
it. I found my footing in the rock-solid farmyard, compressed by the
tires of tractors, combines, and implements for sixty years. I suffered
most of my abrasions there, out front of Mom's fenced-in lawn and
garden, below the grass that grew down over the rise where Grandpa
had built the house. I fell off my bike traversing the ruts, picked up
goathead stickers in my feet and hands, fell off my horses there, and
felt the whack of clods thrown by my brothers.

Other surfaces were home to me as well—my horses' backs, the corrugated tin- and asbestos-shingled roofs of buildings, but there was always the return, the jolt to my legs as I leapt off the pig shed, the wallop of my head or ribs on the ground when I fell. Flame, a sorrel mare just home from the horse breaker's, hurled me off in the farm-yard just below the knoll where my mother's lilacs bloomed. *Wham.* Windlessness, pain, shock. Those things happen to you over and over, but you get back up. You always get back up from the ground, its imprint on you until you die. Then they bury you in it.

Ground had two meanings—the usual stuff we stood on when we weren't inside and the stuff we farmed. Sometimes we just called it dirt. The yard beyond my mother's garden was unimbued space, the given I thudded over without noticing. Toward the fields I felt more respect, if also boredom and oppressiveness. I knew that the crops grown there gave us our livelihood, but they were endless plains to trudge over, my ankles buckling on the soft ridges. The only intriguing thing about the fields were the gullies, where pulverized quartz and mica sparkled in the silty beds. Sometimes I even found arrowhead chips in the gullies.

My grandfather Carlson had traded his Texas dirt for Kansas dirt, sight unseen, then moved his family and found it good. Twenty quarters, we said with pride, because 20 times 160 was 3,200 acres, making us Big Farmers. The crops varied, but winter wheat was the baseline. Dad planted it in September, and it sprouted before the end of the month—mechanistic, perfect green stitches in the dirt. The first big growth came in May, and I would lie down in it, my back in one row, and my arms and legs growing down the adjoining ones. When the heads formed in early June, their whiskers would hiss and sigh over me. Looking into the vortex of blue overhead, I felt myself expanding. Before long, I had to stand up. Even if I was a mile from the house and no one could see me, I needed to be identified somehow, confirmed. Standing above the earth reoriented me and put me back into what seemed like the right relationship with it. The concept of ownership was deeply imbedded in me. Our land belonged to us so thoroughly that I never worried or mused that it might be the other way around.

Of the four classical elements—earth, air, fire, and water—earth, or dirt, is the one I knew most intimately. It mingled with my sweat, and I tasted it when I licked the salt in the crook of my elbow. On Sunday mornings, when my mother and I got into the car to go to church, the wind hurled it, mixed with road gravel, against our legs. Whenever the wind blows in the city where I live now, I remember the low-pitched hum it made as it rattled the casements in the farmhouse windows. I hear my mother's voice, "Doggone wind. Cussed dirt."

Dust was our industry's by-product. We manufactured and dispersed it every time we touched implement to earth. In the summer, my father and my brothers disked the summer fallow to keep the ground free of weeds, and the dust swelled behind the tractors in white clouds. It filtered into my hair, weaseled into my navel and socks. It was the ground airborne.

When people live near rivers, they tell flood stories. For us, it was Dust Bowl stories. My parents and aunts and uncles all had them. The schoolhouse my father attended had grown dark as night during one afternoon blow. That same day, Mom's brother John, who'd been pulling the one-way over wheat stubble just west of the windbreak, was forced to park the tractor and follow the fence line back to the house with his hands. In one of our family photo albums, there's a picture of a dust cloud about to envelop the barn and a corral full of cows.

Whenever I griped about having to do the dusting, my mother would tell me that I should have been there in the Dirty Thirties, when the house filled with dust. "It was an inch-deep around those baseboards," she would say, pointing to the corner of the living room beside the big bay windows. "We used scoop shovels for dustpans!"

I trailed rags and mops saturated with furniture polish the color of red licorice over the house's surfaces, careful, as Mom instructed, to get the crannies. The narrow pine floorboards in the upstairs hallway and bedrooms glistened momentarily, but dulled as soon as the polish dried. Its kerosene scent gradually faded as the dust returned.

The dust rains down on me now as I sit at my desk in this distant place. It's as if I am standing on the front porch steps these decades later, while my ten-year-old self beats the cotton-looped mop against the upstairs balcony rail. The dust never left. It has lingered in my subconscious all this time—dust from disturbed earth, turned and tilled sod, of fields stripped bare then sowed, of western Kansas.

They have TV commercials in Kansas, spawned by the state's tourist board—a waste of money. Kansas is impervious to tourists. Only natives can endure it, and some of them do so reluctantly. But like all places, Kansas has its boosters, ever hopeful. The commercials cut between images of sunflowers, grain elevators, windmills, and wheat fields. A chorus bursts forth with the satisfied pronouncement, "Ah, Kansas!" Well it could as easily be, "Ah, dust!"

Dust was everywhere, in everything. In the rafters of the long, low-lying sheep barn, it smelled oily, like lanolin, and lay in fine, mice-tracked heaps. The smell equaled danger, for the slats my brother and I walked along disappeared at one point, and we had to leap from joist to joist. One room of that building smelled of Hershey—a grain used in birdseed. We jumped from the rafters into it and the reddish yellow beads rolled over our bodies like a billion miniature ball bearings. The Hershey millet felt cool, like water. It didn't itch like barley or oats. They were the worst. Come near a scoopful of barley and you went home with its dust in the waistband of your jeans and under the elastic of your underpants, itching like crazy. Whereas the wheat— *that* we could shower in, standing in the combines' hoppers, letting it pour over us, freshly harvested. Wheat dust powdered our skin like talcum.

Dust in the loft of the big barn smelled of disintegrated straw and pigeon droppings. Downstairs smelled of straw, rolled oats, and horse manure, with variations in the tack room where Dad kept his bagged soybean meal. The east end, where my brothers milked the cow every morning and evening, smelled both of cow and horse. Dust in most of the outbuildings was the dried sweat, afterbirth, saliva, and excrement of animals and the powdered remains of what they ate and slept in. In one building it was primarily pig, in others, sheep, cow, horse, chicken, barn swallow, or pigeon.

These are just the broadest classifications. There were a thousand subtle differences and combinations within each structure. In the shop, I inhaled the smells of various lubricants—hydraulic, drip, gear, and engine oils, grease—and dust. When I dared to peer into the shop's basement, shafts of light from a little south window revealed the jutting blades, springs, knobs, and handles of implements and tools that my grandfather Carlson had abandoned there, now covered with dust. That basement smelled not only dank, but foreboding, and for good reason. More than once I'd seen black widow spiders dangling from the webs they'd strung between the artifacts.

The house had its own characteristic dusty odors. At night, I would lean my elbows on the sill of my bedroom window, my forehead against the screen. A loamy scent lofted off the wet dirt in my mother's flower garden and filtered through the nearer odor of the moth carcasses that had collected in the casement, then turned to dust.

The pastures were another world, miles of dustless grass. One of my brothers pooh-poohed the arrowhead chips I found in the dirt of field gullies and instead looked for entire points on pasture hills. The Indians had liked to camp high up, he said, and the arrowheads you found there still lay where they had fallen, perhaps in battle. When my mother and then my two older brothers attended the nearby one-room school, the teacher used to take the students on field trips to our pasture's "canyon," a craggy outcropping overlooking the S-curves of a dry creek, the Little Beaver. Ten more miles into Cheyenne County, and the terrain became sandier, with more yucca and cactus.

I lived on the edge, I realize now, of the western encroachment, a border region, not precise and mappable, but as it had evolved due to climate and soils, down on the ground in one little corner of one big county. It wasn't a sharply drawn line, but it made itself known by the way—between the Johnsons to the south and Aunt Irene's to the north, and the Seamans' to the east and Grandpa Bair's to the west—pastures gradually began to predominate over fields. These were fenced, yes, and perennially grazed, but they hadn't been scraped clean yet, denuded.

I absorbed the feel of that grass too, its dry and tangled close-napped matting was soft only in color—a subtle, soothing green in the spring turning gradually to straw yellow in the fall. It spanned into the distance, dotted by blossoms of apricot mallow. I trudged across our pasture countless times, rattling a grain bucket and attempting to conceal a halter, trying to catch our horses. Although just plugs, they looked beautiful running over those grass-clothed hills. At Grandpa Bair's, my cousin Judy and I would go into the pasture to sneak licks on the salt block, telling ourselves that we only licked the corner that belonged to the Guernsey milk cow, not any area that the more common Herefords used.

Even though I was on most intimate terms with the disturbed ground, especially the farmyard — lumpy and bumpy, some nails and spent machinery parts embedded here and there, stickers, and dust — the grass beyond stirred me to wonder. In the spring, when our thousand head of sheep milled restlessly in the corrals, awaiting shearing, I sensed that our settlement, with our shoddy outbuildings, our pecking chickens, and digging dogs and children, was an invasion. Agitated and angry, the sheep turned in a giant spiral at the gate leading out into the pasture. The call of the grass reached over the ewes' backs into our farmyard, and into my dreams.

My father didn't bother to keep up the buildings. He put his effort into where it paid, in the fields. My mother had all she could handle with her acre of yard. On rare visits from town friends, I cringed at how run-down the place looked, even though my grandfather had built a grand house by local standards. It sat on a rise, and you could see it all the way from the county gravel, east a mile.

The space on the south side of the house, beyond Mom's front yard, still feels charged in memory, because that's where our road had once continued, going west. I would peer beyond the new corrugated tin prefab garage, past the windbreak, to where the track disappeared in a section of wheat, and all of that road's lost potential would swarm in me.

The road would have been an excellent shortcut to Grandpa Bair's place and the girl cousins who lived there off and on. We may as well have been on ocean islands, so dependent we were on the rarely departing pickups and boaty sedans our parents drove. The road in my imagination also led into a nameless eternity of pastel green buffalo grass, into a past and a future, a frontier. At night in the big house in my solitary bedroom I would dream our wheat field behind the windbreak back to buffalo grass. I would ride my horse on and on, over inclines, where wild animals grazed. There were no fences, just lungfuls of pristine space.

Once, I rode my horse through the wheat field and opened a neighbor's pasture gate, my heart pounding for fear I wouldn't be able to relatch it and would have to confess to my parents that I'd disobeyed their orders to stay on our property. The gate opened onto a fenced lane, no sign of a road, just buffalo grass underfoot, the same as the pastures on either side. After about a half-mile, I came to a dry creek bed too steep to cross, and no way to go around because of the barbed wire. The creek was the north fork of the Little Beaver. It meandered all over the countryside, and in some past flood, or series of them, the bridge and even its pilings had washed away. There was no sign of it anymore, just the smooth, tantalizing sandy bed that I dared not play in. My parents had warned me off the creek beds, because rattlesnakes burrowed in the banks. Danger—this made the missing road to my grandparents' place all the more enticing.

Later, when Grandpa Bair died, the vanished road echoed with a new feeling that I could never quite pinpoint. I know today that it was grief, a connection severed. Not only had the road vanished, but the destination had as well, and we no longer had any way of reaching it even if we took the long way around.

Although hundreds of miles and many years separate me from it, I still know that land, not like the back of my hand, but in my palm —from having lifted it and sifted it, raked it into miniature lanes and fields, from having buried treasures in it, bled into it, cried into it, dreamed upon it. I even know the roads intimately. I dragged my fingers over them as I coasted in the rusted red wagon, a hand-me-down from my brothers. I remember the feel of the water-washed grooves and cracks in the dirt grade going all the way out to the county gravel.

That road had two hills and I usually lingered in the valley between them. I liked the strangeness of not being able to see our house a quarter-mile away, while I could still see the tall white grain elevators in town. Those elevators beckoned magically, because they were "there" not "here," and it seemed I only ever got to be here. The town they sat beside was so far away — twenty miles — and full of kids who, like my cousins on Grandpa Bair's place, I had no power to reach.

Most often the roads were just another place to live the earthbound, immobile life of a child. I would sit down there on the earth that was home, playing, pretending, waiting while adults took too long in the fields or in their boring conversations. My parents were the most confident people you could imagine, descendants of pioneers who had laid claim to the land as far as their eyes could see. Their future was guaranteed by the ground underfoot and by us scrabbling children, who they were certain would carry on.

I, on the other hand, could hardly wait for Saturday morning, shopping day. My mother would back the Impala out of the prefab tin garage and we would race toward the twin grain elevators, dust blooming behind us the whole twenty miles.

THE FIRST WORM

from *Daddy Long Legs — The Natural Education of a Father*

JOHN T. PRICE

*S*pencer charged into the bedroom where I was resting.
"The first worm, Daddy! The first worm of spring!"

He stuck his fleshy, three-and-a-half-year-old fist in front of my face, allowing the mud to drip onto the white sheets. The short end of an earthworm, brown and moist, was attempting to wriggle through the hole between Spencer's palm and pinky finger.

"Wow," I replied, trying to sound enthusiastic, "that's a lively one."

"Don't you want to hold her, Daddy?" He opened his palm to show me the entire worm, more dirt spilling onto the sheets. It was long, but a little thin as far as Iowa earthworms go. Then again, it was only March—by July it might be as thick as my finger.

"Don't you want to see where I found her, Daddy?"

I didn't respond right away, and I could see the disappointment in his face. Normally, I would have put more effort into this moment. Around here, the first worm of the year is the equivalent of the first robin, the first tree buds, the first daffodil shoots. The first definitive sign that winter is nearly behind us, the season of life begun. This worm had appeared a little earlier than usual, making it even more special. At that moment, however, I wasn't very interested in worms. I was trying hard to relax and, as my doctor put it, "take a close look at my life."

Let me back up a few days.

Late in the evening, around the time of the March full moon—the Worm Moon—I was sitting at my desk, trying to write, when it felt like someone reached into my chest, grabbed my heart, and juiced it like an orange. Pain shot through my entire upper body, even my teeth. My first thought was: *Could this be the gas station burritos I scarfed for dinner?* Then it suddenly got worse, doubling me over in the chair, robbing me of breath. I thought I was going to pass out. Gradually the pain diminished and I began to breathe normally.

Definitely the burritos, I rationalized. *Now let's get back to work.*

"Not the burritos," my wife Steph insisted a couple of days later. She demanded I immediately see a doctor.

"That's ridiculous," I said.

"Then I'm calling an ambulance to carry you there. Lights, sirens, the whole shebang—you want that?"

An hour later, I was sitting in the doctor's office while he listened to my chest through a cold stethoscope.

"I don't hear anything suspicious, but we need to run some tests. First, an EKG, then a stress test to check for damage. You may have suffered some kind of stress-related cardiac event, but it could be tough to determine now. You should have come in earlier."

"Do we really have to do all that?" I whined. "I'm not that stressed out—I'm not even teaching right now."

"Actually, it can be very stressful for some men to get away from their normal work schedule, especially workaholic types. How many hours do you typically work a week?"

Silence.

"Trouble sleeping?"

Silence.

"Inexplicable muscle aches and pains?"

Silence.

"Bouts with depression, listlessness, irritability?"

Silence.

"OK then, shall we get on with the tests?"

When I pulled into our driveway after seeing the doctor, Spencer was crouched near the front retaining wall, turning over rocks in his Superman pajamas and red rubber boots. As usual, he was under-dressed for the weather, which still carried a wintry sting.

"I'm looking for worms," Spencer told me as I walked toward the front door. "Wanna help?"

"Not right now, buddy," I said. "Maybe later."

I entered the kitchen, where Steph was waiting. I told her the EKG had been normal, emphasizing again that "the event" had probably been nothing.

"Did he prescribe any medication?"

"Just baby aspirin and some predictable advice about eating healthier and exercising." His actual advice, which I didn't share, was "You need to eat healthier, exercise, and take a close look at your life. If you want to live, that is."

I think Steph knew I wasn't telling her everything.

"Whatever this turns out to be," she said, "it's a serious wakeup call. Why don't you go upstairs and rest for a while."

Lying in bed, I thought about the doctor's ultimatum: *If you want to live.* Of course I wanted to live—I was thirty-nine years old! Stretched out on the bed, eyes closed, I had just begun to scratch at the surface of this troubling thought when Spencer made his entrance with the worm.

"Open your hand, Daddy."

He dumped the creature into my palm along with a pile of dirt. The worm wriggled and twisted, trying to dig its way into the muddy flesh it confused for home.

"Her name is Wilma. Please come see where I found her, Daddy. *Please.*"

"OK," I said, and arose from my soil-splattered shroud.

Outside, Spencer took my hand. We walked past the retaining wall near the house where I'd seen him searching earlier, past the scraggly pear trees and the giant blue spruce, and across the open ground. The yard, like the worm, was not pretty during that time of year. A raggedy half-acre, our sloping, uneven property looked naked in early spring, the veneer of snow having just retreated, exposing brown patches of grass and cold mud. Everything was mashed down and dead-looking. Sticky baklava-like layers of leaves had accumulated against the backyard fence and in various spots where they'd blown and settled last fall. I stepped in one of the gooey piles and got a whiff of decay.

"Hurry up, Daddy!" Spencer commanded, pulling at my arm.

Spencer and I stepped down to the crumbling retaining wall at the far northeast corner of our property, near the woods. With the leaves gone, I could just see Cedar Lawn cemetery on the other side of the trees, its gray and white stones dotting the hillside. This part of the yard was largely uncharted territory for Spencer and had always been a little scary with the cemetery and all, but he was clearly in the process of exploring it up close. I felt proud of his daring.

"This is where I found her," Spencer said, pointing to an overturned rock at the base of the wall. "I'm going to let her go now, so she can see her momma. Her heart's wrinkled because she misses her so much."

Always the momma, I thought.

"And you can't tell anybody where she lives, OK."

"OK," I assured Spencer, "I won't tell anyone where the worm lives."

"Her name is *Wilma*, Daddy."

He crouched down to search for her original hole, but couldn't find it. After digging his finger into the dirt to start a new one, he tried to coax her into it, but she just squirmed around on the surface. Apparently, that spot of dirt wasn't good enough for her anymore. She wanted something better. *Just who does she think she is?* I thought. More than a worm? A metaphor? A squiggly sign of subterranean imbalance? Of mortality—*The worms crawl in, the worms crawl out, the worms play pinochle on my snout?* I looked again at the cemetery on the other side of the trees, and wondered: If I'm buried there, perhaps sooner than expected, will Wilma's offspring find me and recognize one of their own? A fellow creature who'd once wiggled and crawled and spelled words?

Spencer gave up looking for Wilma's original hole and turned over another rock. There was an unexpected gathering of roly-polies and a small centipede, as well as the glittering strata of another half-submerged earthworm. Their appearance in our yard, like Wilma's, was earlier than usual. I immediately thought of global warming, but not Spencer. His eyes widened—the cold earth appeared to be giving birth—and he started touching the bugs with his fingers, chasing the centipede into the brown grass and scaring the roly-polies into becoming little balls he could roll around like marbles. Talk about stress.

It seemed like a good time for me to escape—I had some serious thinking to do, after all. I quietly made my way to the kitchen door, but before going inside, I took another look back at Spencer. He hadn't noticed my absence. He was in the process of turning over yet another rock, then jumping aside to avoid crushing his toes. He crouched down close to the earth, his red boots standing out starkly against the bland winter colors. I couldn't tell what new life he was discovering there. Only that I was missing it.

THE LANGUAGE OF CLAY

❋

ROXANNE SWENTZELL

*M*aybe I shouldn't have eaten so much dirt as a baby or licked the adobe walls of our house so often, but dirt tastes good. Pueblo children that grow up in the old villages know which houses taste the best.

Then it was our mother's clay . . . *ummmm*. Mixing dry clay and water to lay out, thickening it to form vessels, I licked my fingers like it was cake batter. Letting the wet clay dry on my hands, I watched while it cracked. My skin looked a hundred years old at age ten! So cool!

No wonder dirt became my voice, my first language. Words couldn't make sense the way clay could. It molded around my thoughts and feelings, perfectly transcribing them into human forms. I found a language out of the dirt that went far beyond words. I sculpted what was happening to me in clay figurines. The clay picked up the slightest turn of a head or glance of an eye that held information all humans understood. It would take a novel to explain what those gestures meant. This clay language goes directly to the source of our human souls.

As I sit with my aging self, I reflect back on this journey of mine and smile at the way I've been loved by this Earth. She has fed me, clothed me, and sheltered me, and she gave me a way to communicate with others. My hands are still cracked with dried clay, either from the dirt in the gardens I work at saving seed and growing out food, or from the houses I build from the adobe dirt that fills the valley where we live, or the clay I sculpt into another verse of my life's story —to tell anyone who might be listening with ears and hearts turned toward this lovely dirt beneath our feet.

You clay people
who dance through my soul
dance right on through me.
My eyes look upon you
out there — I know you
in here. Like children
out in the world
I send you
and hope
you find love
out there.

DIRT

Imago Ignota

*

JOHN KEEBLE

*M*y earliest memories of dirt come from when I was a young boy of four. We lived on a hill and during springtime I would combine the dirt with small stones and sticks and construct experimental earthworks to guide the water of the snowmelt into little lakes and dams. Sometimes a small stick would double as a boat, enter a rivulet, and careen downwards. I suppose it was mud I played with, dirt mixed with water. There were the mud puddles, too, the bane of mothers and a great source of pleasure for young people in galoshes who were fortunate enough to have dirt around them. I first lived in a small town in Saskatchewan and we had plenty of dirt in those days, all right.

There was the dirt I remember when I was not much older, a patch of it near the steps at the back of our house. I sat on the steps in the sunlight with a stick in my hands and drew in the dirt. I was brought to consider infinity, as I had lately been struck by the meaning of "The End," and then by the question of what comes next after "The End." This simple, rudimentary contradiction was a childish insight into the nature of things, and while my phrasing of the question has grown much more ornate, I can't say that my understanding of its meaning has improved in the least. What strikes me as fascinating is that I was drawing a figure with my stick in the dirt while trying my hardest to unravel this matter. The question seemed to emanate from the dirt, radiating through the squared-off head and querulous expression of the figure I had drawn. It said something about what I might see now as the classic fundament of elements . . . earth, water, air, and fire . . . but which then was merely the grounding sense of touch with a solid thing, holding the stick in my two small hands, touching one end of the stick to the dirt, and moving it to outline the rudimentary head while my mind went off toward the empyreal, sparking the imagination. It was an obscurity felt as an inchoacy, an "imago ignota," and it is important to consider the order in which those two things came: first the grounding, and second the sparking.

When I was eight, my family moved to Berkeley, California, where my father attended the Pacific School of Religion. There on the lavishly planted and somewhat unkempt grounds of the campus I found myself transfixed by a slope overgrown with dense bushes surrounding a single, huge fir tree, which I watched during storms from our apartment window. The tree tilted, bent, and whipped in the wind. One spring day, I made my way to the tree by crawling beneath the bushes. Upon reaching it I found a tremendous gnarled root system clutching at the dirt. The brown needles that fell from the tree made a thick duff, eventually to be transformed into more dirt, and there were spiderwebs that held entrapped flies and a colony of sow bugs, which curled up into balls when touched. Those things were the grounding there. It was a potential, frangible detritus, found in a dark place, and, I thought at the time, safer than any other place I knew of, solitary and secret. I had to creep out the way I came, emerging covered with dirt and with cuts from the thorns and brambles.

My family moved to Southern California, where I had a transfigurative experience with dirt. We traveled to Death Valley to camp for several Thanksgivings, a time of year when the desert was cool. We went with friends of my parents, the Sayles, for whom I recall having great affection, though now I know little about them, except that they were artifact collectors and old enough to be my grandparents. They had no children. We camped in a place with a hot springs, which was near what seemed a vast plain stretching as far as the eye could see, but with few plants growing on it. Instead, it was littered in places with small stones of agate, jasper, flint, opal, and obsidian that had been chipped by human hands. It was a stone-flaking ground and we would walk along, traversing the flat with our heads down as we searched for artifacts, and I remember one I saw . . . a pink-colored piece of opalized agate. I bent to dig it from the dirt. It seemed presentational, an ensconced arrow point, and I can envision it still, the dirt framing the luminous stone. I lifted it to show it to Mr. Sayles, who touched his finger to the fine flaking on the point's gently curved hafting and pronounced it to be a two-thousand-year-old ceremonial or mortuary point.

Whether he was right or wrong, I have no idea. I do not possess the arrow point and I think now it was possibly ceremonial because of the ornateness of the hefting, but probably not two thousand years old. At the time, I knew nothing of the value of artifacts, and certainly I did not consider that the original makers, feasibly Panamint Shoshone, might wish to lay claim to them, and thus that what I was engaged in was a form of plunder. Though at age ten or eleven, I was at a time when my consciousness was dividing into what some hold as the signal stage of growing out of childhood, the nagual (familiarity with the non-ordinary) giving ground to the tonal (a fixation on the ordinary, the everyday), the possibility that what we were collecting came from a burial ground did not register, perhaps simply because it was not a part of the conversation among the people there . . . the Sayles, my parents, and myself. It would take Barry Lopez years later to articulate that for me. In his essay "The Stone Horse" he describes his encounter with a horse made of an outline of stones, a four-hundred-year-old intaglio laid into sunburnt and sandblasted desert varnish, which is a patina of iron and magnesium oxides. He says, "[T]he few who crowbar rock art off the desert's walls, who dig up graves, who punish the ground that holds intaglios, are people who devour history. Their self-centered scorn, their disrespect for ideas and images beyond their ken, create the awful atmosphere of loose ends in which totalitarianism thrives, in which the past is merely curious or wrong. . . . [But] I remembered that history, a history like this one, which ran deeper than Mexico, deeper than the Spanish, was a kind of medicine."

From the experience of finding the arrow point in the dirt, misunderstood as it was, I developed one sometimes useful habit, that of searching the ground when I walked, of being alert to what the dirt offered up, to the sparking that helped me make my way as an adult. This is how I begin to write my books—the sparking point for new material. This is how I walk the Eastern Washington farm where my wife and I have raised our children and lived for forty years: deer carcasses, cow carcasses, a heifer practically disemboweled by her breech-born calf, all manner of carcasses going into the ground, raccoons, porcupines, mice, and gophers, flies and maggots eating the dissolving flesh, dust from taking the care to disk in the residual organic matter left after baling hay. It's garden dirt made into soil, I suppose, and it is not unlike writing novels. Chickens and dirt, cow manure and dirt, deer manure, convoluted moose droppings and dirt, snow and dirt, rain and dirt, dirt from dirt roads, dirt in the nostrils, in the cracks of skin, imbedded under fingernails, dirt storms, veritable clouds of dirt, great plumes of dirt blowing across the Pacific from the Far East, for nothing is strictly local.

There was the stratospheric column of volcanic ash from Mount Saint Helens that covered our place in 1980 and floated around the world, this and more on ground covered by one of the largest floods known in the history of the world, more than twelve thousand years ago. The ice dams broke apart at the end of the last glacial age and the resulting floods inundated hundreds of square miles from Missoula, Montana, to the Pacific Ocean, covering portions of Idaho and Oregon and much of Washington. Where we live there are fields where the once massive eddies slammed into the hills and turned, dropping their loads of dirt. Within sight of such fields there is basalt on the surface, known as scab rock, where the water raged, washing the dirt away.

It is interesting how the word *dirt* has undergone a nearly 180-degree turn in meaning in our culture. The word is thought to emerge originally from Old Norse . . . *drit* . . . meaning excrement. *The Oxford English Dictionary* lists this first, "1. Excrement," but there are other meanings too. "2. Unclean matter, such as soils and any object adhering to it; filth, especially the wet mud and mire of the ground, consisting of earth and waste matter mixed with water. 3. Mud, soil; earth; mould; brick-earth," and it adds the more metaphorical meaning, "4. The quality of being dirty or foul; dirtiness; foulness; uncleanness in diction or speech." I have a copy of Webster's dictionary dated 1911, which defines dirt as: "1. Any foul or filthy substance; excrement; earth; mud; mire; dust; whatever, adhering to anything, renders it foul or unclean."

The change seems to have happened sometime in the last century.

In 1927, Hermann Hesse published his novel *Steppenwolf* in its German edition, and near the beginning of the book he has his willfully shabby, dirty, and unkempt protagonist, Harry Haller, pass by a place "so shiningly clean, so dusted and polished and scoured so inviolably clean that it positively glitters. . . . Don't you smell it, too, a fragrance given off by the odor of floor polished and a faint whiff of turpentine together with the mahogany and the washed leaves of the plants— the very essence of bourgeois cleanliness, of neatness and meticulousness, of duty and devotion shown in little things? I don't know who lives here but behind that glazed door there must be a paradise of cleanliness and spotless mediocrity, of ordered ways, a touching and anxious devotion to life's little habits and tasks." Haller goes on to claim . . . undoubtedly ironically . . . that he is not being ironic.

The tension Harry Haller foresaw is out in the open now, for the world's population, and its inventions, have increased exponentially while the earth's dirt in its frenzied and fecund form has proportionately diminished. We've also come to understand very well the new dimension added to dirt. No longer is dirt always a thing that needs to be washed out like Macbeth's "spot," made hygienic and sanitized. One does not think solely of one of a number of secretive, contagious killers, cholera, say, looping around a village in the mud and water, E. coli poisoning from animal waste stirred into fields of salad greens, or instruments with viral contaminations, such as HIV. On the contrary, we've come lately to think we've grown excessively clean, that our immune systems require more contact with the minerals and myriad of microorganisms, which, if one were to dig one's hands in the dirt, one would come up hefting a load of the visible and invisible . . . earthworms, larvae, tiny insects, tiny snails, nematodes, and bacteria, frangible fossil matter, frangible sticks and leaves, carbon, and radioactive isotopes, some of which might contain the germs of yet unrealized cures. My wife and I have a dog that eats dirt every springtime and a grandchild who adventures in it just as I once did, searching out his inchoacy and the seemingly random sparking. If only we could cease our plundering habits, the products of human invention that strain to drive the earth into utter exhaustion. We're playing a fool's game with our dirt, blindly transforming it "behind that glazed door . . . [into] a paradise of cleanliness and spotless mediocrity" through genetically modified crops and monoculture, herbicides, fertilizers, coal mining, petroleum extraction, and fracking, that dire, unthought out, and "awful atmosphere of loose ends in which totalitarianism thrives."

MUD PIES

*

CHRIS LARSON

*S*heer delight is mostly what goes into a mud pie.
   Almost no one has ever eaten one, at least not a whole one.
A childlike energy rises with the gritty coolness of the memories;
   small hands open and intimate.
Then, I heard **IT** clearly.
IT said: *You have got to be kidding!*
And the child vanished.

And left me curious.
Can something that addresses me as "you" really be me?
I watched as small hands made those mud pies ten times more.
You are too old!
You are too busy!
When are you going to GROW UP!
The mud will never come out from under your fingernails!
You don't even have any pie tins, FOR GOD'S SAKE!
(I think **IT**'s missing the point, don't you?)
What if someone sees you?
What will they think?
You'll lose your job!
Do you actually LIKE shopping carts and dumpsters?!!!
You take the prize for stupid ideas!
And the child vanished ten times more.

Such dedication to stealing the sparkle!
IT talks to me like no one I call friend.
Will I ever smooth the crown of a mud pie again if I have to convince
   IT to come along?

If **IT** were really on my side, at some point, wouldn't **IT** be more . . .
   well, *kind?*

Each moment is a last time for something.
A last kiss, a last word, a last full moon.
And I can't help but wonder about those mud pies.
If I asked the sky or a tree or a meadow or a dandelion, what might
   they say?
You can if you want to!
The messier the better!
We thought you'd never ask!
Let's do it together!

Curious about just one thing now.
If I never listen to **IT** again, how would that be?
YOU CAN'T DO THAT, YOU'LL DIE!!!

Oh, I see.
Well, I'd love to stay and chat,
but each moment is a first time for something
and there are these mud pies waiting for me.

SERVICES AT THE CHURCH OF DIRT

❋

MARILYN KRYSL

*A*s a kid I liked dirt's soft, powdery feel between my toes, and I could dust things with it—earth's talcum. Summer days I wandered my grandparents' wheat farm, tossing up small tornados of dust in the corral, then sitting and sifting this fine stuff through my fingers. Stunned by heat those long afternoons, I stirred water into a pan of dirt. So unlike each other, dirt and water, but together they became a third thing: mud. Dirt and water were the primal goo, the basic material of the universe. Stirring mud suggested food, and the combine's flat lengths of tin looked like cookie sheets. I laid out my delectables and left them, baking.

One day I mixed up a pan of mud and made an Adam and Eve. She had breasts, he had a penis. I remembered that they were innocent, they couldn't see each other's nakedness. So I gave them no eyes. Their mouths cried out in the wilderness of dark, so I mixed up a pan of brownies. When they'd baked, I fed my creations. And while they ate I began the long contemplation.

The Bible said dirt was what people were made of. *From dust thou comest, to dust thou shalt return.* Everybody came originally from dirt. Every *thing* came from it. Dirt was everywhere, and it was what the earth itself was made of. Dirt was precious stuff, our basic material. Elemental, magical.

I began to construct a philosophy of dirt. Dirt was not dirty. It was a sacred powder, sacred stuff, and that made the earth sacred ground. Holy. I piled dirt in a cracked blue willow bowl, set it on top of a box, lit a candle beside it. I knelt before this altar, murmuring devotional incantations. The air agreed with my procedures. Water running underground seconded my motions. I had made a shrine, and in the church of dirt I held services. At the close I stood and sang *America the Beautiful.*

Digging graves was good. We buried the dead, because it was fitting to put them back where they'd come from. When my Adam and Eve began to crumble, I scooped out a hole and laid them in, covered them. I stood over this funeral with the hose a long time, watering. Adam and Eve were losing themselves, becoming the ground they'd come from.

Gardening was good for dirt. Fat worms drilled their passageways through topsoil, giving the roots of plants an airing. And loosening the top of the ground with a spade let the earth breathe. When I turned earth over with a trowel and crumbled it in my fingers, it fluffed up. Dirt liked being handled, as I liked being hugged.

I talked to dirt, told it stories and sang it songs. *Do unto others as you would have them do unto you.* It seemed self-evident that we owed the ground every consideration. I had doubts about tractors though. They made a terrible noise and their blades sliced. I was not in favor of building dams, and I didn't approve of oil well drilling either. The dynamiting of perfectly good hillsides upset me. I didn't think men driving bulldozers should smash down fine stands of wild grass and weeds, leaving behind the imprint of tire treads.

"Why are they doing that?" I asked my grandmother.

"They're just building another road," she'd said.

How many roads did it take to get from one place to another? One, I thought, was plenty.

Stories of the Gold Rush made the miners sound like men gone mad with greed. But even before I heard those stories I felt uneasy about mining. I didn't like to imagine picks hacking away at a mountainside. The rock could feel it. Rock was sentient, not quite in the same way animals and plants were sentient, but in its own way, rock's way.

I felt in the features of the landscape a resonance. Earth's electromagnetic web? I felt vibration. I was a sounding body for the Schumann resonance. I stood amidst the trilling presence of my surroundings. I was alive amidst other live presences. I registered the sentience of lakes, hillsides, and pastures, and I worried. We were hurting things, I thought.

My grandparents had a mulberry tree and on good days I climbed this tree and looked out across distance. I swayed on my branch like a girl at the top of a mast, and I thought I could see the curvature of the earth. I could feel it, way out there, starting to curve down. Anything that big, I thought, had its own plan. It wouldn't do to try to move things around on the globe as if they were furniture. Let things just lie there the way they are, I thought. If the earth wanted to move things, it could make an earthquake or a volcano or a flood. If the earth made a cave of the winds or cut out a channel for a river, it knew what it was doing. Those things were earth's business, I thought. Who were we to interfere?

There was nothing to do with these feelings of mine. No one I knew shared them. Eventually I learned not to voice them. But at the time, in my solitary moments, I registered earth's irritation and wondered. When no one was watching I held a rock to my ear and listened, calling the universe for news.

And sometimes I knelt and laid my ear against the ground. I heard, far down, a grumbling, like thunder in the distance, but round, a spherical grumbling, vast in its dimensions, a sound as large as earth itself. It was, I thought, the moan of a creature, the audible irritation of a great being. Smohalla the Dreamer said, "The white man tells me *plow the ground*. Shall I take a knife and tear my mother's breast? The white man tells me *quarry stone*. Shall I dig beneath her skin for bones?"

I'd been right, I thought. We shouldn't be bothering dirt with bulldozer and dynamite. Like anyone, the earth needed soothing. We ought to be patting and smoothing, watering and blanketing. We ought to be talking and singing to and reassuring our ground.

I did not know, as a child, the true extent to which the earth had reason for irritation. I mixed dirt and water and was oblivious of the Trinity test, Hiroshima and Nagasaki, the testing of the hydrogen bomb in the Marshall Islands.

Now I think what I had then was a wonderfully irrational wisdom. Ancient people felt metals had a life of their own, like animals. Stone was the concentrated power of earth, and precious stones were frozen nectar, the life force crystallized. Thales, a Greek physicist, believed chunks of amber were alive. Native Americans know mountains are living beings, that wind caves are a mountains' lungs. These lungs breathe, and this breath can be recorded. We have measured the breath of caves at the base of the San Francisco Peaks in Arizona. They inhale and exhale at 30 mph in six-hour cycles. Claude Kuwanijuma, a Hopi, said, "The stones remember. If you know how to listen they will tell you many things."

Physics has noticed that the basic bits of the universe are not truly particulate. What is going on in and around us is not the still presence of chunks of inanimate matter, but a process. As Gary Zukav writes in *The Dancing Wu Li Masters*, the basic "unit" of the universe is not matter but an event.

As John Archibald Wheeler declared at the 1981 Nobel Conference, ours is a participatory universe. What there is is the whole, *that-which-is*. We are all of the whole, bound into each other and into the world. Playing rock telephone, it seems, wasn't so silly after all. My notion that dirt liked watering on hot days was a notion wiser than I knew. I lacked knowledge, but I had wisdom, and I was paying attention, not with a child's fantasizing, but in an important, vital way.

Waterfall and volcano, parrot and pickerel were like me, and I was like them. I could deduce from the way I felt how fine dirt felt getting sprinkled, how much grass liked being left alone to flourish. How birds fluffed up with well-being when they were sung to, how crickets and beetles profited from storytelling. How the branches of the mulberry tree appreciated bodily contact as much as I did. How light needed my prayerful attention. And how darkness rested in my sleep.

Wise attendance is more than disinterested observation, as we are more than collectors of data. We are part of the data. We are sentient creatures, of the mesh, of the whole, and we can't step outside of ourselves because we are what we attend. When we help or hurt what only seems to be outside us, we help or hurt ourselves.

Sentient beings, we wander the earth, drink the water, eat the fruit, sit on the ground. Knowledge of our ignorance brings us down to ground level. Ground level is human level, sentient level. Sitting in the dirt, without instruments, we are sentient creatures. Sitting in the dirt, without instruments, we begin to see.

We see mite and sprout, feel heat and light. We hear wind and water. We pick up handfuls of dirt, we begin to wonder. Sentient, we notice we're here, in and of the universe. From this perspective, everything is as important as everything else. Sitting in the dirt, we are at last in a position to become philosophers. Finally we understand deforestation hurts the forest. We know the granite upthrust registers the blast.

I would mix again some dirt, some water. I would fashion another Eve, another Adam. I would give them genitals again, those necessary parts. But this time I would also give them eyes. Let them sit on the ground and gaze at each other. They compare, they notice the differences between them. And they notice as well the similarities. Both have hair, both have legs, arms, fingers, faces. A nose, a mouth. Ears. And those eyes.

When they have looked their fill at each other they notice where they are. They notice the earth beneath them. Eve picks up a handful of dirt, feels it. Adam picks up a rock, puts it to his ear. Eve's fingers register the texture of dirt, its fineness. In sunlight the dirt looks bright. Suddenly, a sound. *Was* that a sound, over there where the leaves moved? Eve pokes Adam. He takes the rock from his ear, looks where she's pointing. Leaves. A bird?

Sunlight, a breeze, insects, flowering shrubs, grasses. These things are a felt field. There is something in the air, an animation, an expectancy. The clearing where they sit has a life of its own, their own.

They look around, they listen. Now Adam too picks up a handful of dirt. There's something here, he thinks, something it seems important to notice. He pours his handful of dirt into Eve's hand. They examine this dirt, they notice in it some quickening energy. Isn't it bright, this dirt, and alive inside its texture. They feel the life stuff sparking in it as they sift it through their fingers. They begin to notice they are not alone.

3

DIRT WORSHIP

that motherly feeling

KIDS AND RUGBY PLAYERS AND COWGIRLS
ALL ROLL IN THE DIRT — THEY'RE JUST TRYING TO
GET THAT MOTHERLY FEELING ON THEM.

LISA JONES

DREAMING IN DIRT

*

BK LOREN

I wake sometimes having dreamt of dirt. All night long stories rise up from vast expanses of gritty soil, landscapes that read, to me, like hieroglyphics, like love notes, like survival tales, like novels. When I look down at my feet, the earth becomes a pop-up book, a 3D world surrounding me with animals I know well, and others I've never met personally, but whose stories I read like the tales I loved as a kid, the ones where people walked with wild beasts — rabbits, deer, foxes, mountain lions — and we all shared a language, because we do. It's the language of dirt.

Every life leaves an imprint. Imagine the earth, then, like a fist that holds your actions and the actions of every living thing. When you learn to read dirt, you walk into the forest or across a city (yes, there are imprints there, too), and the fisted world opens up like two palms holding a book of the best story ever told, because it is every story ever told — if you know how to read dirt.

When I see dirt in my dreams, I know I'm going to have a wild sleep. A scratch in the dirt tells the tale of the survival of some small animal, a mouse, a mole, sometimes even a house cat, and the same scratch tells the tale of a wild hunter that went hungry after working hard for a meal.

Even the story of flight imprints in dirt. Where a magpie lands and touches the tips of its wings to the soil, where an owl swoops down and leaves five parallel arcs that embrace a sharp divot where talons extended and clenched. Sometimes the talons lead to the imprint of small feet that darted out of the night hunter's grasp. The escape looks like commas on a page, like rainfall gone sideways on the earth. Sometimes you can see only a large almost-circle, like a period inside the parentheses of the owl's wings. That's the end of the sentence for the prey, no dashes telling of its escape.

Sentences become paragraphs become novels written in dirt.

When I wake from a dream of dirt, I know I need to slow down, to look more carefully, as I do when I'm tracking. That's the lesson of dirt. There are lives we step over daily. When we walk in a wild place, we are never alone. Lives cross paths with ours in every moment. Almost every time you hike in Colorado's Front Range, for instance, you're watched by mountain lions or foxes; or you're studied by coyotes, those tricksters who see you — and then you watch them disappear like smoke that sinks into the earth rather than rising into the heavens. You cross paths with animals that walk side-by-side with you like some kind of post-Edenic apparitions, reminding you that the only original sin is one of not paying attention, of not listening to the dirt beneath your feet, the signs all around you, the names of the animals fresh on your tongue, their stories waiting to be spoken alongside your own.

When you walk, you know we are all woven together in a texture of earth that transcends the boundaries of language. If I walk where my parents walked, they, too, appear again, not in their own tracks, but in visible signs: the two Canada geese they called "The Couple," one with his distinctive limp (unnoticeable until you look at his tracks), and the other with the dot on the right side of her bill; the small pond Mom and Dad walked around daily (hand-in-hand at seventy-five); and the tracks — and springtime eggs — of the Western painted turtle they saved from a busy road and brought here, where it has thrived long after their deaths. It's dirt that holds my parents now, dirt that will hold the ones we have loved and their imprints forever: fossilized bones, mummified shapes caught in the act of living and dying. It's dirt that will hold you and me.

Dirt is everywhere and records everything, retelling your story, perhaps even eons after your death, in sediments pressed into history, pressed into time. There is nothing you do that escapes record. There is nothing that the earth will not record and read back to you and others. Listen: it's ever-present. Our lives left in dust, where our stories, always, remain.

TAO OF DIRT

✳

LIZ STEPHENS

I am crouching on the hiking path above this major highway, poking at coyote scat. Poop, obviously. What passes between my daughter and me is wordless only because we've said it so many times before, she can see what I'm showing her: all the fur in the scat, how tangly it looks, how dusty and undense and near to dirt it already is. Other people hike by while we crouch there and if anything looks more strange than a mother and daughter poking at animal poop together, it must be the two of them doing it so silently.

She thwacks it as long as she wants with a stick while I stand to stretch my legs and gaze at the highway. When I look down, before we resume our walk, I see that she's hurried the work of weather along and has effectively dispersed the scat to parts: eyelash-tiny pieces of fur, threads of sticks, and for all intents and purposes, dirt.

In the desert, behind the house of a friend, we stand the way the houses do—perched impermanently there, cool shells against the force of the land and weather outside—and sift through owl pellets. Poop again. Owls live in these palms above us and the ground here is covered in science, for anyone willing to brave the wrath of the bees that have also settled in these palms. Such a verdant five square feet. We start slow, my daughter and I, picking lightly at the pellets as our hosts make conversation with us. Then as they and my husband wander away for cleaner pastimes, we dig in completely, ripping the pellets open bare-handed with our thumbs, rasping them in our hands to rub the fur and sticks and bones apart into components, little stacks of like things. We make lines of perfect mouse skulls, piles of tiny ribs. We sift the broken work out into the quickening air of evening, fulfilling the last step of the cycle as all that life falls as dirt again. We lean close and inhale and rub hands on jeans and generally do all manner of things we should not do if these owls or mice were the carriers of any disease at all. We stop just short of rubbing our eyes or mouths with our hands, but that's the only rule I make ever dirtiness-wise. When we traveled in the subways in New York City, I found myself shooting my hand out against all odds of timing to come between her shoulders and any wall, her cheeks and any windows or seats, sliding in at the last moment to cradle her individual self from contact with such a mysterious place, thousands of layers of selves. But in the natural world I am the opposite. No doubt I'll get in trouble for this, and I do joke that when a human inevitably carries the next bird influenza or monkey germ to the people, it's certainly coming from us.

It is with my hands I draw the world to me and dig into its workings; I'm greedy for it. I watch other people watch things, and then hold my breath as they pass and then grab it, and then feel it, and put it to my face. Seaweed and dead crabs on the shore, bark of a tree with bugs, the lost feathers other parents rip from their children's hands — "germs!" — I want those. I take them and then I give them to my germy child.

The only sickness we ever get is from other people.

Dirt, it turns out, has *Mycobacterium vaccae* in it. In part, that's Latin for cow dung. Poop again; the first place the bacteria was isolated was in cow ("vacca") dung, but if it grows there, as we know, it is from and on the way back to dirt. Recent studies indicate that *Mycobacterium vaccae* initiates the creation of the chemicals within us that function as antidepressants. This surprises me absolutely zero.

I garden. So in that way I make life out of the dirt, out of what the dirt can do; I dig my hands into the world and make it interactive. But it's not the planting of seeds that one gardens for; it's the tending of lives. I fret with light and water. I am Prospero to one bit of land. Still the difference between my casual interest in gardening and the depth of my connection to the way life plays out on the ground is the difference between enjoying choir music and going to the church for the sermons; one relaxes me, but the other is my lesson, my lifeline, part of my pilgrimage.

The way dirt accommodates death is what I'm after.

I would like to be buried, I think, intact in whatever cardboard box might last just long enough to get decently through the memorial service and into the dirt. And if the next day it rains, all the better.

I used to want to be cremated. The thought of the bugs that so thoroughly finish the work that birds of prey and wolves begin on dead wildlife horrified me. I don't think any of us have to scratch too deep to access an image from somewhere of a worm crawling out of an eye socket. But now that I have held nature so close to my face, in my hands, and watched it so long, I want what the other animals get. I want to be scattered. I want to be carried far places. I want to go into fox dens, wend my way through the earth, run over the ground into reaches I can't reach now, and fly.

Who doesn't want to fly?

And then, finally, I want to become part of the dirt. Because it's not going anywhere but over the face of the earth, over and over again.

My mother carried a jar of red Oklahoma dirt everywhere she moved when she left there. I deeply understand that impulse. But I feel like the animals bring the dirt of the world to me. On their feet and coats, but also in their nature, so close to the ground, so dependent on the conditions of weather and elements, so brief in life span between dirt and dirt, so cyclical and anonymous in their strife that the fact that they arise from the dirt and live at all seems the miracle.

And we are animals too, of course. We are built in such a way that we must face the earth by choice (our gaze is not on the earth, and so we face it only by choosing to) but we just aren't very far from it. We clothe ourselves against it, we house ourselves away from it, and may not touch it directly for weeks or years, but gravity assures that we are rarely more than inches away. And we are going back to it. So that something, someone, else, may use it next.

THE LIFE OF SOIL

*

BERND HEINRICH

*I*t was black. I remember, because at my age then—around eight years old and in early spring—it was the first time I had looked down to notice dirt. Our family was quartered in a one-room hut in a dark forest in northern Germany right after the war. Towering pines shaded the ground except for a small clear patch up on a bank in front of the cabin.

Light snow had recently covered the ground, and now, after a warm spring rain, it had become black, and that made me notice something marvelous by our doorstep. Quite suddenly, maybe one day to the next, I saw a small patch of the dark dirt turning luminous green, and it was larger the next day, and then expanded in ribbons over the black ground from one day to the next: I was mesmerized by a rosette of laterally spreading grass blades that had made the dirt vibrant.

This had been, as far as I can remember, my first moment of wonder. Had grass been underfoot before, I would probably have hardly noticed it from seeing it all the time, nor would I have made connections from it to dirt. But the single patch that had expanded from one day to the next had transformed this dark dirt to what I might later have called Life. It was a moment of magic and mystery, maybe even of ecstasy, or it would not have been stamped into my memory.

For the time being then, dirt was also something crumbly under the soles of my feet and between my toes. It was the sand on a mile or so of the woods road between our hut and the village school. Shiny green beetles flushed in front of me, and after a brief zigzagging flight, where they glinted like jewels in the sun, they landed a few yards ahead. We called them "sand beetles," and later tiger beetles. Although I couldn't fly, I could run, and it felt good to be on par with such gorgeous company.

Tiger beetles (of the family Cicindelidae) are related to carabids, which are commonly called "ground beetles." Like me, ground beetles do not fly, but they all run. These earthbound running beetles soon became my passion, to have and to hold, through the influence of my father. I accompanied him when he dug pits in the ground for catching mice and shrews, but ground beetles fell in, too. He gave me a field guide to identify them, and I would know them by name, the giant black *Carabus coriacus*, dark bluish *C. intricatus*, shiny copper *C. cancellatus* and its look-alike *concolor*, and the deep green *C. auratus*. . . . The merit of these intricately sculpted beetles was not only that they were beautiful, but also that I could find them by scanning the ground wherever I walked, and I could also catch them.

I thought of my old carabids with a start, with a nostalgic recognition, when recently—now in Maine, on a new continent—I dug out the pit for my privy. There, several feet in the dirt, I unearthed a carabus. There, in the chaotic substrate of unimaginable complexity that looked like sheer or mere randomness on my shovel, I suddenly found one of those creations of the dirt. It was metallic black, sculpted in lines and pits, and its edges glistened a deep purple. I did not know the name of this species, nor what it was doing there, but I kept it in a photograph. Perhaps its larva had burrowed and metamorphosed there to become an adult, or maybe it had hibernated there in the winter, or escaped heat or drought. But in any case, it was ultimately from and of the dirt, just as the green, green grass had been. The beetle had likely fed on snails, and the snails on grass. It was of the soil, which I was preparing to receive my wastes. And this same receptive soil would also receive all of me eventually, to again convert all to grass, trees, and flowers. For the time being, though, an American chestnut tree I had planted there years earlier, as well as the nearby sugar maples, would grow well because they were next to the privy.

I used the dirt from the pit excavation to make a raised garden bed, into which I planted potatoes. I just stuck several of them into this dirt, and presto, come fall—it seemed too good to be true—there were perfect and delicious Yukon Golds.

My girlfriend, Lynn, saw the magic, and before I knew it we had an even bigger bed of potatoes, beans climbing a pole, snap peas growing on chicken wire, and little green sprouts of kale and carrots. We had watched with eager participation as the emerging green dots in the dark dirt first turned into shoots, and we would harvest potatoes in August for eating in winter. With every meal there would be a memory of that plot and our shared task. There is more to be had from dirt than food.

I think Thoreau knew this well and maybe said it better. A hundred-seventy-two years ago, Old Henry was "determined to know beans," and having made himself a two-and-a-half-acre bean field, he tended and hoed it daily from "five o'clock in the morning till noon." He came to "love" and "cherish" his beans and said, "They attached me to the earth and so I got strength like Antaeus." He became, as he said, "much more intimate with my beans than usual." Along the way he concluded that "labor of the hands, even when pursued to the verge of drudgery, is perhaps never the worst form of idleness." And he told the reasons why.

When tending his bean field, Thoreau was "attracted by the passage of wild pigeons," he "sometimes watched a pair of hen-hawks," heard the brown thrasher sing, and with his hoe "turned up a sluggish portentous and outlandish spotted salamander." His enterprise was "not that I wanted beans to eat," nor was it likely for "leaving a pecuniary profit."

I'm in rapport with his romantic ideal and his enjoyment when pausing "to lean on my hoe, these sounds and sights I heard and saw anywhere in the row, a part of the inexhaustible entertainment which the country offers" — as opposed, I suppose, to those summer days "some of my contemporaries devoted to the fine arts in Boston and Rome" as entertainment instead. Perhaps this vibrant "idleness" is what Thoreau cherished most, and maybe freedom.

It may seem ironic that, with time, I felt at war with grass, as Thoreau did as well. It threatened our beans, and we longed to see the dirt exposed. Dirt is wild. It nurtures everything, and our labors make the difference.

One would, however, want to get real when it comes to dirt and work. We do not generally hoe beans in order to hear the brown thrasher, or exhume a spotted salamander as lofty ends in themselves. One wants to see the things impartially from all perspectives.

True to that sentiment, however, Thoreau gave an exact economic enumeration of his work as well. He itemized pecuniary costs and profits, in which overall bean patch costs added up in his accounting to $14.72 and ½ cent, with a profit of $8.71 and ½ cent.

To our ears now, Old Henry pretty much worked that summer in his two-and-a-half-acre bean patch for nothing. The "bean (and peas, beets, potatoes, etc.) patch" that Lynn and I are working on this summer allows for some comparisons. We are not seeing any passenger pigeons, but we are getting similar pleasures to what Henry got from his. Plus we enjoy companionship, which Old Henry did not appear to pursue. So for us it's a win/win situation with the dirt, in more ways than two. But I also suspect this dirt will, before the start of next winter, also become a winning economic activity as well. And so was Henry's (if he'd excuse me for being familiar), despite what he may have implied and what we inferred.

Our dirt patch is 1,600 square feet (0.037 acres), so his was about seventy times larger. He spent $3.12 on seed, and we spent $94.00. Thus, overall, in terms of *our* money, he paid about one-thirtieth as much overall, but on a per acre basis, in dollar amount, he paid all of 1/2,100th as much. Take outside labor: his "ploughing/harrowing/farrowing" cost him "$7.50." (This amount irked him, because in *Walden* he added a comment — "Too much" — for emphasis next to it.)

How much is his "Too much"? Lynn and I paid our neighbor, Mike Pratt, $150 to harrow our plot, which as already mentioned, is one-seventieth as large as Henry's. But Henry did not pay seventy times as much. Instead, he paid one-twentieth as much overall, which comes to 1/1,400th per acre of what we paid. Similarly, our total pecuniary costs were 1,960 times more than his, prorated per acre. My point is that inflation since the time of Thoreau's bean patch (of 172 years ago) has reduced the worth of a dollar to about 1/2,000th of what it used to be. Thus, Thoreau's seemingly trivial profit of $8.71 and a half cent is actually a hefty $17,430 in terms of the dollar now. (And his seeming pettiness, accounting to the last half cent, is thus more like figuring to within ten dollars now.)

How many young people today could earn $17,000 in a summer

by working forenoons in a bean field and having the rest of the day off for "other affairs?" None! But it was not the amount of time Thoreau spent in his bean field, or the money he made, that he rhapsodized about. It was the ancillary "profit." Now we are hard-put to get a fraction of the pecuniary profit he earned, and even then it is usually at a cost of the main ancillary ones that a country life close to the dirt provides and that we now all too commonly lack. Thoreau derided husbandry as he saw it then, pursued with "irreverent haste and heedlessness . . . [with] our object being to have large farms and large crops merely." His conclusion that "I will not plant beans and corn with so much industry another summer" suggests he felt that even his "industry" was already too much.

To turn now to the other Henry, a Maine writer a century later near the arguable beginning of industrial agriculture, who marks the end of dirt with its ancillary gratuities and the beginning of accounting by and to the dollar its immediate worth only. In his book *Northern Farm*, Henry Beston reminds us "the shadow of any man is but for a time cast upon the grass of any field. What remains is the earth, the mother of life." And he concludes, "When farming becomes only utilitarian, something perishes — sometimes it is the human beings who practice this economy, and often most of all it is a destruction of both land and man."

As a fellow Mainer, and as a fellow human — united not by artificial or perceived boundaries but linked together by our universal bonds to the soil of Earth, the link that connects all of Life — I grow beans for more than utilitarian purposes. My farming may be token, but like the blades of grass that first sparked my interest in living things, the activity is a visceral and vital reminder of the grandeur in our very existence.

DIRT IN LOVE

*

BARBARA RICHARDSON

I first see Lele in the garden with a machete. No, I hear him
before seeing him. Chopping corn husks, chucking them in the
compost bin, wearing a stained white tank shirt and no hat, body
sweating in the midday August heat. Though in his sixties, Lele is
so strong he does not let up with the blade even when he sees me.
He doesn't spare the energy to smile. He nods. And when I do not
fade behind a row of towering pole beans interlaced with cucumbers,
when I dare to stay for a friendly tour, Lele's eyes cast about for Thea.
Thea's the tour guide. Thea is his ambassador of dirt.

Lele (Raffaele) Malferrari stands six feet tall with a grizzled pony-
tail. Tanned. Silent. His *work* is his words. Thea loves him dearly. Her
bright eyes tell it. In fact, when Thea talks about dirt or pottery or gardening or Lele, she is the most beautiful woman I know. She laughs
and croons a tale or two riding alongside me in her three-wheeled
cart through their enormous garden, ticking off her favorite harvests.
Snapping beans between her teeth. Making me sample the bounty of
Thea and Lele's well-tended dirt.

Clay, soil, mud, slip, it's all dirt and it's all worthy. Take their hand-
built adobe house—which required seven years to finish. One year
in, when Thea faced the dirty block walls cemented with dirt mor-
tar, heaps of dirt and dust and broken guesswork, and a fine green
haze of seeds sprouting through all the grouted seams, she nearly
wept. Although now when I say "a Chia pet home!" she laughs about
it. They've lived in their adobe house thirteen years. The seams
smoothed by hand and the high walls burnished with creamy mineral
paint—cool in summer, warm in winter. Built to embrace a shaded
courtyard that opens onto grape arches giving entry to the community garden, backed by a chicken house that is nearly as large as their
red pottery studio just beyond it. I am reminded of Suzuki Roshi's

"one whole world," no inner, no outer. Thea and Lele have everything they need in their rural commune on the dry plains north of Boulder, Colorado.

It is easy for a visitor to admire.

Lele pauses to watch a scrawny rooster dip over the rim of his sturdy pallet-and-straw-walled compost bin. "Renegade," he says with approval. Lele's constructed a penthouse of eggshells and avocado pits and a new guest just checked in.

Meanwhile, Thea extols the virtues of Borlotti beans and double-dug planting beds, buckwheat crops planted solely for bees, and the valerian just pushing skyward to make their biodynamic tea, a sort of homeopathic love brew that Lele will spread with a spray backpack and wand over the whole foliar-fed garden like a blessing every two weeks. What is blessed grows best. As she leans into the handlebars of her sturdy trike, Thea's ardor for all things green lilts through the garden like bees.

I know with body certainty that I am witnessing dirt in love.

Then Thea tosses her head and stretches her arms up to crown her soft gray hair, remembering how she and Lele met, and years fall away. Her eyes ignite. She assures me dirt love started with her father, a Dante scholar and professor who taught simply for the time off in summer to garden. And what a garden it was! But at age twenty-nine, when Thea took a break from art school in London and traveled to Italy to see her sister, she met a dashing young Italian — a blue-collar mechanic with luscious red-streaked locks so long everyone mistook him for a musician. Lele worked steel, designing templates for factory goods. He didn't spend his summers working the dirt or winters dreaming about organic gardening, like her father. He was a tall, inventive city boy, an urban anarchist who never even graduated from high school. Thea plunged into the sweet summer romance knowing that in the end, she'd go straight back to finish art school and then home to her boyfriend in the States.

But Thea was miserable back in America. She dreamed of Lele day and night, night and day. Her boyfriend seemed all wrong. When her body began sending distress signals, she visited her doctor. His diagnosis was MS. His advice to Thea Tenenbaum was simple: *Go out and do what you want to do.* She remembered the bold pink roses Lele had sent her to mark her graduation from art school. The next day, Thea bought a plane ticket to Italy. This newly diagnosed, garden-loving art potter plopped herself down in an Italian stranger's care, saying, "Here I am."

Thea left behind not only her life and kin, but her beloved art form: stoneware. Italy has no clay suitable for throwing stoneware pots. Italy has terra cotta, a porous earthy clay. It fires at low temperatures and behaves nothing like the elegant mud Thea had used in London. She does not complain. She tells Lele exactly what she needs for a pottery studio. And tells him she'll teach him, so he can quit the factory. The factory where Lele has seniority and a pension. All his friends and family shake their heads like he has gone insane.

Some would call Thea impulsive. Some would ask, *What does dirt in love do?*

Dirt in love searches the countryside and finds an old wooden cart wheel with a metal rim, a giant round slab, and hauls it home to fashion it into a potter's wheel for its clay-hungry beloved. Then builds a kiln based on pillow talk and scavenged bricks and penciled diagrams. Then dirt in love (aka Lele) scavenges the environs like a hawk for the very clay—red here, green there, gray, brown, white—that his mate will fashion into nests of many sizes, cup nests, bowl nests, that they will paint and glaze and fire.

The process is exhausting.

Of course the glazes are utterly different being slurried from Italian soil. And Lele refuses to work the wheel, just digs in his heels at Thea's promptings. Fortuitous refusal. Thea loves throwing pots. And all of Lele's years in school doodling in notebook margins leap to life in the skunks and frogs and whimsical birds he paints around Thea's promising nests. Thea soaks Lele's rock-hard clays in a bathtub in the barn. Clay he's hauled home in his backpack. Clay he's stacked inside their car. Clay from the mountains, clay from an old medieval pottery site nearby, clay from the river—so much mixing and experimenting. In the *stalla*, the cow stall, an amply pregnant Thea spends hours bent over that bathtub forcing wet slurry through a sieve. Removing tiny shells, the calcium that absorbs water and explodes in a kiln.

Homemade wheel. Homemade kiln. Handmade pots. Glazes that bubble or vanish. Thea and Lele endure so many failures they line their driveway with broken pot shards that say, *This is the way to dirt in love's house.* Their newborn son Emiliano absorbs the trials and errors. When they reach the States to settle in at the commune in Boulder, his classmates dub him Michelangelo. Glass, metal, clay, wood, bread making, tagliatelle making — the boy can do anything with his hands.

In time, Lele adapts to country living, commuting to work via garden rows out to the pottery. Accompanied by chickens. Thea embraces it all, the work, the family, the fresh eggs, the earthy fertility — her father's daughter. And her loving impulse toward life, her daring lifelong artistic vision, hold firm through the rigors of the slowly escalating MS. Her right leg numbs and stiffens. Walking is difficult. Still, Thea works the garden rows at dawn while the dew lies heavy. That's when the lettuce picks fresh. She serves hand-plucked salads in immense painted bowls. She gathers friends for family feasts under the morning glory arbor. I have eaten those salads and supped on her stews, under the spell of Thea's feisty radiance. The light in her increases every year. Diane Ackerman says, "Look at your feet. You are standing in the sky." That is Thea Tenenbaum. Dreaming dreams so rooted in soil, heaven and earth connect and shine.

Now in their sixties, Thea and Lele have created a dirt-based business that did indeed take Lele away from the factory, away from the urban Italian life he loved, to the open-hearted city of Boulder, where they built a dirt-brick house, raised two kids, turned a goat-milking shed into a working pottery studio, produced a magical line of terra cotta ceramics, and took charge of the enormous communal garden that graces their current home. They have tended this soil for thirty years. A half hour down the country road, their daughter Gelsemina tends sheep, gardens organically, and teaches outdoor skills to city kids. She just gave birth. To little Lutreo. Thea's brown eyes fill with tears. Another gardener. *Just look at those strong hands.*

Lele is quick to say, "I am not a gardener." Some people "have the calling" but Lele insists he does not. He has a partner who loves gardens, so he chops debris with a machete in 90-degree heat. If you ever have the good fortune to taste the bean stew Thea concocts from that garden, the tenderest soft golden Borlotti beans melting into a vegetable base too wondrous to be called minestrone, you might understand Lele's devotion. "Il Maestro" his neighbors call him. The maestro of soil.

But Lele and Thea are both masters. Both hold the first principle of great gardening to be *no guilt*. "If you feel guilty, it's better that you stay at home because we don't want the guilt in the garden," Lele says. "It should be a pleasure. And that was the hardest part." Harder than transforming the shallow clay soil into rich brown fragrant loam, harder than double digging every last bed, harder than weeding, and constructing a thirty-foot-long tomato hothouse, and engineering rows of seven-foot-tall pole supports for the climbers year after year. I walk around this garden guilt free.

With dusk approaching, Thea pedals her three-wheeled bike from their house to the beds to pick dinner — the sweetest trip to the grocery I've ever seen. Lele yanks a bouquet of tall green dill as a farewell gift and puts it in my hands. Then we squat at the small new serviceberry shrub he planted last year at my bidding. I'm a landscape designer, with a fondness for fruit-bearing natives. Five dark berries have eluded the birds. He eats two, I eat three.

I grip the dill bouquet. I say, "I love it that you're here." He says the same to me.

This is what dirt talks about. It is always in love.

DIRT HOUSE

❋

PETER HELLER

I know a little bit about dirt. I have a dirt floor in my house. Not kidding: three layers of poured mud, cracked out, patched with mud again so it looks like flagstone, treated with mineral spirits and linseed oil. It has the texture of rough-out leather and looks like a night sky if the night sky were kind of muddy: the bits of chopped straw I mixed in with the mud are flecks of gold like sprays of stars if stars were partially made of old hay. A mud/hay universe.

Also, the little chest-high wall between the kitchen and the wood-stove is made of dirt. It steps down in a flowing cascade, one-two-three steps like a desert mesa. Dirt bricks that capture and hold the heat from the stove and radiate it back all night. You can't tell, you might think the wall is made from 2 x 6s because it is covered with stucco, cream with a touch of pink, like the inside of an old adobe church, but nope, it is real dirt under there. On two of the steps are wood decoys of loons, black and white with little muddy red eyes. Yep, eyes like mud. The decoys look like they are swimming over a waterfall.

The wall in back of the kitchen, and interior walls of the two bed-rooms, the deep window sills, and the outside walls that are eighteen inches thick? They are all dirt. And I'm the son of a gun who built them.

After living in this house on and off for many years, I wonder why everyone in Colorado, in the whole dry West, doesn't live in a dirt house. Because a dirt house with windows on the south side never freezes. No matter what.

The house is built into a low hill in Paonia, Colorado. It's a lovely valley, a winding of orchards and vineyards and ranches along the North Fork of the Gunnison all nestled down in the mountains like a train set, with the West Elk Mountains to the south, Grand Mesa to the north, the Raggeds and Beckwiths to the east, and desert mesas and plateaus stepping out to the west. It captured me in a way no other place ever has. I wanted to build a house here and settle in. I was a freelance writer and a poet and didn't have a ton of money, and I traveled a lot so I needed a house that I could leave happily without worrying about freeze. That's when a man came to town with this machine. It was yellow. It had a big hopper. You dumped dirt into the hopper, a hydraulic ram pressed the dirt at 2,300 psi into a steel box, and out onto a track of steel rollers kicked a pressed dirt block, eight inches by sixteen by five. "Green," meaning the bricks are not true adobe, which are poured into forms and laid out in the sun to cure and harden. These are not cured but pretty tough nonetheless. The man's recipe called for mixing sand and chopped straw into the dirt as binder, but we found that the clay soil here was full of little flakes of a harder shale that worked perfectly.

I had Bill Pitt come with his bulldozer and dig a flat pad out of the low hill and pile the excavation dirt. Then he spread a foot of gravel. My carpenter friend Don helped me build a footer. In a wide straw hat like an impressionist painter would wear, and chewing over a mouthful of sunflower seeds, Don built my door frames and lintels and a stack of window bucks out of local, standing dead, red fir, and we erected the door frames on the footer so they stood like empty glyphs against the mountains behind them, like some kind of ironic Stonehenge for an empty age.

So here's the cool thing about building with this brick machine: we dumped in the dirt from the hill, straight, unadulterated, and the machine kicked out the blocks. The crew and I stacked them in a double wall on the footer, right around the door frames, and we set the window bucks on the blocks where they offered the best views—to the south—"Hey guys, nudge it over a little to the left, yes, there's the top of the mountain, good!"—and we built the brick walls around the windows, and the house grew up from its origins and the dirt piled back in to where it had settled thousands of years ago, to where it belonged.

We stuccoed it over and put on a shed roof, and the sun streams in the big south windows, and the thick walls and the dirt floor absorb all that heat, and even on the coldest winter nights with no fire day after day, the house will never get below about forty degrees. When it's sunny for a few days the warmth rises, and when you light a fire it gets extremely cozy. And the cool thing, or warm thing, is that it's a much different kind of heat. It radiates from the floor and the walls. You can carry groceries inside on a bitter night and leave the French doors wide open for a few minutes and the house will keep radiating warmth from all sides. It is soothing being inside this house. It's all curved windowsills and curved corners, because you can do this with stucco, so there aren't many real edges. Also, I read that dirt has pheromones, or something, that come out of the ground and mix with our endocrine systems and give us a sense of well-being. In this way dirt is like potatoes and tobacco and opium.

The house has a very low profile. It looks like it belongs there. The back wall, the side set into the hill, is only four feet high, and once a cow elk hopped up onto the roof just to look around.

The elk straining the rafters didn't scare me, but water does. If my dirt house ever gets wet enough it will turn back to mud. That almost happened last year. Wet late winter, wet spring, the ground in back got saturated, and the bottom two feet of the back wall, soaked and muddy, began to bulge. Scary. Jeff Roop came out with a track hoe and a congenital inability to get ruffled; he dug the dirt from the whole outside back wall and dug a trench along the footer and put in a French drain and now we are okay.

The house is off the grid. It has four sixty-watt panels, which are plenty. A propane tank in back fires an on-demand water heater and powers the fridge and gas stove. It is not net zero, but the dirt cabin consumes very little energy. And it just feels good when you're inside it, like sitting in a big lap. A friend from Hollywood came out recently, an active screenwriter and director who is in the middle of that whirlwind. He settled into the flannel sheets in his dirt bedroom and slept for eleven hours straight, something he said he hadn't done for maybe twenty years. It may have been the flannel, but I'm betting on the dirt.

I just had a memory from my grad school days in Iowa City. It was early fall. It had rained hard for a few days and a group of twenty of us met to play touch football. We stood around on the sodden grass and chose sides as it began to drizzle again, and someone yelled, "Let's make it tackle, what the hell." Holy shit. You went up for a pass and were tackled and slid for ten feet. You barreled into a runner and grabbed him and others piled on and churned a brown slop that spattered up into your face and you collapsed into a gritty wallow. In half an hour everyone was mud head to toe. Nobody could stop laughing. When we were all numb with cold and shaking—Were our lips blue? You couldn't tell—we broke up and walked home, leaving mud footprints on the sidewalk. It was one of the happiest days of my life. Is there some connection to the quiet joy I feel in my house? Some dirt paranormal phenomenon? I don't know. But give me a mud/hay universe any day of the week. Give me thick walls of dirt to lean on, and a floor that looks like the tail end of a mudslide, with constellations of gold stars to walk over.

SINKING DOWN INTO HEAVEN

*

JEANNE ROGERS

I am a Midwest farmer's daughter and as such no stranger to dirt—450 acres of it to be exact. In addition to growing sweet corn, field corn, alfalfa, oats, and wheat, we raised dairy cows, beef cattle, pigs, and sheep. The dairy barn took up the western section of land next to the woods. The steer barn, corncrib, and hayloft claimed the eastern border near the creek. The sheep grazed on the northern edge close to the swimming hole and the pigs wallowed in the mud to the south. My bike and I were constant travelers on the gravel roads that connected the respective barns and outbuildings, and I'd be lying if I said that navigating those loose gravel roads on my black Schwinn with skinny tires was not a tricky endeavor that required Band-Aids and Mercurochrome on a regular basis. I rode my bike from one end of the farm to the other, and when my legs grew tired, I high-tailed it over to the swimming hole. I ask you: what could be better than a hot dusty bike ride followed by a cool swim and a lazy sunbath? Often, as I lay atop my dry clothes, I imagined the earth spinning faster and faster, imagined myself clinging to the warm grasses for dear life so that the centrifugal force would not spin me off into the ethers. Other than my grandma's house, there was no place I would have rather spent a summer afternoon.

When the sun neared the grove of trees in the west, I headed home to our well-worn, two-story white farmhouse, its wide front porch and spacious yard shaded by oak trees. Our family garden plot, sixty feet by thirty feet of fecund soil, ran alongside the fence that separated our side yard from the cow pasture, and in it we grew everything imaginable: raspberries, strawberries, cantaloupe, tomatoes, cucumber, yellow squash, zucchini, acorn squash, green beans, sweet peas, lettuce, spinach, broccoli, rhubarb, green onions, white onions, carrots, potatoes, peppers and the best sweet corn known to man. Many afternoons found me on the front porch staring off toward the neighbor's property on the far side of the creek while I snipped beans, shelled peas, and daydreamed about the swimming hole, a permissible destination only after chores were completed. Much like the sunrise and sunset, chores began and ended each day.

My duties consisted of tending our garden—weeding, hoeing, thinning, picking, and preparing the fruits and vegetables for farm-style, midday meals and evening meals we called supper. As farmers, we didn't use the word dinner for the evening meal. To us, dinner meant Sunday dinner at noon. I never heard the word used in the context of an evening meal until I went to school, learned to read, and discovered that Dick and Jane ate dinner in the evening. That's also when I discovered that town people were not exactly like us country folk. But I digress. Preparing one noon meal and one supper each day for six hungry people didn't begin to put a dent in all of the produce, so we spent numerous hot summer days canning, freezing, and pickling the abundance.

Hundreds of cucumbers transformed into dill and sweet pickles. We put up row after row of canned tomatoes and peaches, the latter bought from a peach farmer who lived down the county line road apiece. In the cool, dank cellar on shelves that lined the canning room, we placed clear Mason jars filled with round, peeled red tomatoes and peeled yellow peach halves, their crimson insides prime candidates for a Cézanne still life. We froze green beans, squash, raspberries, strawberries, and tender sweet corn kernels that we carefully and laboriously removed from the cob using a metal and wooden device that looked like a medieval torture rack. We froze whole strawberry-rhubarb pies and put up jar after jar of my personal favorites: raspberry and strawberry preserves.

I loved every part of the dirt, manure, and water that went into creating our prolific garden. I also loved the dirt, manure, and water that caked on the soles of my bare feet, which were often so dirty that they looked as if I were wearing short brown boots. I was nine years old the summer those dirty feet helped me rise from picking the low-to-the-ground strawberries, stretch my back from having been bent over for so long, place my hands on my hips, and wonder: *How can town people be happy without loving a piece of land?*

That evening at the supper table I posed that question to my father, a man who had more than once shared his regret about not completing college as had his two older brothers, who now wore suits and worked in cities. From time to time he had also wondered aloud if he had made the right decision in remaining a farmer instead of finding a more sophisticated occupation. As he pondered my question, his normally steel blue eyes turned a bright blue and his jaw popped as he chewed — sure signs that he was getting riled up.

"You know," he said, "I was a kid during the Great Depression. Town people lost their jobs. No money. No food. Nothing. Those scrawny town boys pulled up to our farm hungry. And guess what?" He stopped talking. A slow grin spread across his face. "We gave 'em food. For once they needed something from us. That's what having land means."

In the autumn of my sixth grade school year when I was ten years old, my grandma died, and because I adored her more than life itself, everything changed for me that year, including, and especially, my relationship with dirt and the land. True to those times and that place where she lived in southern Illinois, in a small town cradled between the Mississippi River to the west and the Wabash River to the east, there near the Kaskaskia River in Grandma's small town, my father and his siblings placed her open casket in the parlor beside the piano for a wake that lasted three nights and three days. It was my people's time-honored manner in which to pay respects and say farewell. It gave us a few days to become accustomed to the idea of her no longer being with us. It gave us time to be with her body a little longer, time to say goodbye. For three days running, every time I passed by the parlor I glimpsed Grandma lying there in her dark blue church dress. A few times I could have sworn I heard her playing the piano and singing one of her favorite hymns. One time I even thought I heard her familiar chuckle followed by her dentures clacking as she said, *Oh Jeanne — you do beat all.*

On the fourth day they moved Grandma's casket to the church, but it wasn't until after the memorial service when the pallbearers closed the casket that the realization hit me: I would never see her again. We followed the hearse to the cemetery and as we stood beside the open grave, the thought of Grandma being trapped underneath six feet of dirt made me feel crazy with rage. I became hysterical. I screamed, cried, kicked, and carried on something fierce, all to no avail. Nothing could change the ordering of that day. Despite my protests, Grandma's coffin was lowered down into the earth and covered with shovelfuls of dirt, which to my ten-year-old way of thinking had completely and utterly betrayed me. I crossed my arms over my chest and declared my relationship with dirt and the land finished, forever.

Fifteen years later as I watched my two preschoolers play in the backyard, John Prine's newly released "Please Don't Bury Me" came on the radio. The lyrics made me feel as though he shared my aversion to the practice of burial. On that afternoon, I adopted Prine's contagious melody and goofball lyrics as my theme song regarding the thought of being six feet under for eternity.

For Mother's Day, my daughters' preschool teacher sent home a yellow rose plant, and when its blossoms began to fall onto the kitchen countertop, I planted it haphazardly in a sunny spot in the front yard. To my surprise it flourished. By summer's end, after several enthusiastic days of planting other young rose starts, we had a burgeoning rose garden—reds, apricots, yellows, pinks, and whites. For the first time in many years I felt great pleasure as I pushed the shovel down into the earth and inhaled the smell of moist, lush soil. I took off my gloves, rested one knee on the ground, and lingered, my bare hands carefully arranging the soil around the base of each plant, tending to their needs much like I cared for my young children.

The years passed, my daughters left for college, and as I moved from one state to another and from one house to the next, I became obsessed with annuals and perennials. Without consciously planning to do so, in the yards of the new houses I recreated my grandmother's flower garden: pink climbing roses, purple butterfly bushes, catmint, lime green hydrangeas, lavender, yellow day lilies, red carpet roses, white snapdragons, and multicolored hollyhocks. Ushered in by the beauty of the roses, my passion for dirt and its works had returned. But Prine's catchy tune remained my theme song regarding burial; I doubted that would ever change.

Coming face to face with death as an adult gave me the unexpected gift of freedom. Life handed me a three-year crash course during which I lost two close family members and discovered a cancerous lump in my breast. Surgery, followed by seven weeks of radiation that turned my breast an angry, painful red, gave me ample time to ponder my mortality and last wishes. Oddly enough, after living in close proximity with death for three years, I no longer feared it. Death and I had taken time to get to know one another. I felt at peace knowing that I, like my two family members, would one day return, in some capacity, to the earth. My loved ones chose cremation. My uncle's ashes were sprinkled from the deck of a boat into San Francisco Bay. My favorite cousin's ashes were sprinkled in a meadow off a California back road near Lake Tahoe. For my own going-away party, I decided I wanted "Please Don't Bury Me" played, and even though I've always imagined my ashes being sprinkled into the Pacific Ocean from a beach on the Oregon Coast, a different possibility came to me not long ago.

On a hike near my home a vast field of blue camas lilies stretched out before me. Have you ever seen their blue tips swaying in a morning breeze? The sea of periwinkle was divided only by a narrow dirt path. It wasn't a tall mountain that I traveled to. No need for hiking boots or rappelling ropes. The blue field did not appear on a postcard you would mail home from your hotel saying, *This is where we visited today.* No one sold jewelry, photos, hot dogs, or candy, not even — as you will probably be surprised to hear — expensive bottled water. I did not need a guide, so safe was my passing there. From the main road traveled by cars, I simply walked down the narrow dirt path through the blue lilies, every now and again feeling the moisture of the marshland rise up around my feet. How I loved that oozing up and over the sides of my shoes. How I loved that feeling of sinking down — not dangerously down, mind you — but sinking down just far enough to know that I too was planted, or could be, if I stayed long enough, in that patch of marshland dirt. How I loved that sinking down on the flat dirt path into blue heaven.

4

DIRT FACTS

interesting secrets to reveal

ONLY RARELY HAVE WE STOOD BACK
AND CELEBRATED OUR SOILS AS SOMETHING
BEAUTIFUL, AND PERHAPS EVEN MYSTERIOUS.
FOR WHAT OTHER NATURAL BODY, WORLDWIDE
IN ITS DISTRIBUTION, HAS SO MANY
INTERESTING SECRETS TO REVEAL
TO THE PATIENT OBSERVER?

LES MOLLOY
THE LIVING MANTLE:
SOILS IN THE NEW ZEALAND LANDSCAPE

THE SOIL'S BREATH

*

TYLER VOLK

*S*oil seems like passive stuff when I ramble over it during wood-land searches for birds. But when I sit at my computer, assembling data at the planetary scale, the soil reveals itself as one of the most active organs in the Earth's "body."

Organisms living on and within the soil—beetles, worms, and other invertebrate creatures, along with fungi, roots, bacteria and other microbes—produce a ceaseless flow of carbon dioxide as they respire. This flood of colorless and odorless gas, the soil's breath, enters the atmosphere and annually exceeds, by about six times, the amount of carbon dioxide emitted by all human activities, including the burning of fossil fuels.

For thousands of years, before there were factories, before vast tracts of forest were cleared or burned to grow crops and graze herds of cattle, the various flows of the global carbon cycle were closely balanced. The amount of carbon dioxide that passed from ocean to atmosphere matched that from atmosphere to ocean; the carbon from atmospheric carbon dioxide incorporated into plant tissue during photosynthesis was matched by a return flow during respiration by bacteria, animals, and fungi.

But now—thanks to the industries, homes, and cars that spew out carbon dioxide as a combustion byproduct of their appetites for fossil fuels—the scales have been tipped, and the atmosphere's store of this potent greenhouse gas has been growing. This new carbon dioxide then enters convoluted journeys, spreading through the byways of the global carbon cycle, the constant circulation of the carbon atoms that are essential to life on Earth. The result is a net growth in the atmosphere's pool of CO_2, with additions of that new CO_2 throughout Earth's biosphere.

The great exhalation sent from soil to air is key to comprehending the role of soil in the global cycle of carbon. Unlike the simple forms of carbon in the atmosphere (the three-atom carbon dioxide, for instance) and in the ocean (primarily the five-atom bicarbonate ion), most carbon in the soil, derived from living matter, is complex, bound into large molecules (long chains of cellulose, massive blobs of protein). Taken together, these partly decomposed, jettisoned tissues of life, called humus, are interwoven into the variegated quilt of soil, along with bits of minerals, tiny organisms, gases, and water. The carbon in humus is what makes soil dark, crumbly, and spongy. Humus, too, gives soil its luscious "earth" aroma.

Soil has so far largely resisted scientific efforts to decipher much of its inner dynamics, but we do know that almost all the store of carbon in the soil is found within the top three feet (exceptions are the deep carbon stores of tundra and peat bogs). Furthermore, about a third of the gaseous carbon emitted from soil comes from its uppermost layer of decomposing litter. This litter—fallen leaves and twigs, older, overgrown stems of moss, and rotting seed casings—is mined by fungi, worms, bacteria, and other denizens of the surface soil. These organisms and their predators metabolically burn the carbon contained in the high-energy molecules of litter and release it, now linked with oxygen, as carbon dioxide.

The remaining two-thirds of the soil's flux of gaseous carbon is exhaled at deeper levels and must make its way up through the tiny spaces between mineral grains and humus particles. About half comes from the respiration of roots (and from mycorrhizae, the ubiquitous fungi intimately associated with roots). Unlike cells in leaves, root cells do not use up carbon dioxide in photosynthesis; instead, like breathing animals, they are overall emitters of the gas. This root respiration is one reason that gases within the soil of a midlatitude wheat field in summer contain a concentration of carbon dioxide more than a hundred times that of the atmosphere.

Respiration by soil microbes generates the other major portion of the soil's "deep" breath. These tiny organisms feed on the carbon in organic matter, some of which reaches them when decaying surface litter is worked downward by worms, beetles, and other soil dwellers. But the largest source of this high-energy organic carbon is already deep within the soil in the form of roots and their associated fungi. Small roots and fine root hairs in particular are continually withering away as local pockets of soil become depleted of moisture or nutrients. As roots decompose, carbon directly enters into the soil at many depths. We expect to find deep roots in forests, but even grasslands support a pool of soil carbon that may exceed in depth the upward stretch of the tall bodies of the grasses.

Whatever the source of the soil carbon, many microbes spend their short lifetimes transforming it into carbon dioxide. A lot of these humus-digesting microbes have not been named or, indeed, even isolated by scientists because many types cannot be kept alive outside the complex matrix of associations in their soil habitats.

Taken together, the exhalations of microbes, roots, fungi, earthworms, and other organisms generate one side of the "budget" for the soil's carbon pool. What about the other side? In general, for a pool to maintain itself, outbound flows must be balanced by those inbound. Root respiration, for example, depends upon the valuable products of photosynthesis sent downward through stems and trunks. Similarly, the microbes and other organisms at the surface and deeper in the soil are supplied by photosynthesis too, in a more roundabout way, with litter that falls and roots that die. Altogether, the sixty billion tons of carbon in the carbon dioxide vented from the soil each year are nearly equal to the annual amount of carbon newly incorporated into the tissues of land plants globally. (Estimates place the products of photosynthesis consumed by above-ground herbivores at less than 10 percent of the total.)

How do the incoming flux from photosynthesis and the outgoing soil's breath affect the amount of carbon stored in the soil? Earthwide, the amount of carbon in humus is two to three times the amount of carbon in atmospheric carbon dioxide. The pool's mass, however, varies from place to place by more than tenfold. Why?

Generally, the more you eat, the wider your girth. And all else being equal, a soil fed more organic matter by its plants will contain more soil carbon. But all else is rarely equal. Some people remain skinny as rails no matter how many milk shakes they slurp, while others remain obese even on highly reduced diets. The difference is in the rate of metabolism, a difference that can determine the bulk of carbon in soils, too.

The rate at which soils consume the carbon received from plant tissues depends upon the metabolism of the soil microbes, the critters that contribute most to the soil's breath. Their soil's environmental conditions determine their metabolic rates, and chief among these conditions is temperature. When warmed, the microbes step up the pace that transforms soil carbon into carbon dioxide, which then flushes into the air.

A vivid example of the effect of temperature on this flushing is the difference in the soil carbon of tropical grasslands, or savannas, and that of temperate grasslands. On average, helped by a more intense sun, the savannas grow about 50 percent more material than do grasslands in the higher latitudes (45 grams of carbon per square foot per year compared with 30 grams). But savannas have roughly only a fifth the soil carbon of temperate grasslands (400 grams per square foot versus 2,000 grams). The reason for this striking difference is that at high latitudes, microbial activity slows to a near standstill during the cold winter, reducing the exhalation of soil carbon. The resultant bigger pool of carbon remaining in the ground contributes to the richness and fertility we admire in prairie soils.

Human activities, especially since the invention of the plow, also affect the balance between the inflow of fresh organic materials to the soil and the outflow of carbon dioxide, and thus the mass of organic soil matter. As a farmer tills the fields, the soil tends to become warmer and more aerated, which increases the rate of decomposition. On average, soils brought under cultivation lose about a fourth of their carbon pool before settling into a newly steady state. Careful management, on the other hand, can increase carbon retention. Gardeners, for instance, often more than compensate for tilling by adding compost and manure. The increasing numbers of no-till farmers will also help on the large scale. Personally, I'd love to see a global movement to "bring back the soil's carbon."

Large-scale shifts in the soil's breath are under scrutiny as indicators — and perhaps amplifiers — of global change. One research hot spot is the high-latitude tundra soil of Alaska, with its large carbon stores. We anticipate that this trend of rising temperatures will continue as a result of the increasing carbon dioxide. Higher temperatures will certainly invigorate the activities of soil microbes. Particularly in the tundra, soil will exhale more CO_2 and also more methane, which, molecule to molecule, is a more potent greenhouse gas than CO_2. Careful monitoring of the tundra must continue as assiduously as a doctor keeping track of a heart patient's rhythm.

However, net changes in any region's soil carbon pool—wherever that pool is located—depend not only on environmental and biological factors inside the soil. Also crucial are changes in the incoming flux of organic carbon. For most plants, more carbon dioxide in the air stimulates photosynthesis by boosting the pressure that drives the gas into the leaves, where it is used to produce organic matter such as starch, cellulose, and protein. World crop yields are already benefiting to some degree. Both with crops and with plants in the wild, every species will respond differently, some adjusting better than others to changing circumstances. Depending on their responses to increased CO_2, some wild species may drop out completely and others may expand their range to become new members of plant communities. To maintain plants we love in the places we love them, we might be faced with the choice to become global gardeners.

Scientists studying the carbon cycle have been calculating just how much carbon to adjust in their budget sheets for the atmosphere, specifically for these biological feedbacks. Estimates thus far tally that, very roughly, the increased soil's breath attributable today to rising global temperatures might just be balanced by an increased flux to storage in such forms as tree trunks, ground litter, and soil humus. Things are expected to change more and more dramatically with time, and most predictions are for increasing amounts of positive feedback over the coming decades, meaning extra soil CO_2 going to the atmosphere. So when I leave the woods and turn to my computer, my mind boggles at the work ahead: to predict the future of atmospheric carbon dioxide, I must round up deforestation, reforestation, entry and exit zones of carbon dioxide to and from the ocean, the dissolution of minerals, energy technologies, and the politics of international responsibility. But these are still other stories in the ongoing saga of the carbon cycle. For now, I will simply exhale.

EARTHMOVER

❋

LISA KNOPP

After the first spring rain, the earthworms are back: thickening and thinning pieces of spaghetti strewn on the sidewalk; wriggling, pink tubes borne in a rivulet along the curb; smashed smears in the street. Not that the earthworms haven't been there all along. During the winter they burrow deeply and hibernate. But when it rains, legions of them come to the surface where I can see them. I used to believe that they surfaced because the water flushed them out of their burrows. Now I know that they surface before or after a rain because then they can move easily over the surface without drying out.

Once the rain stops and the earth begins to dry, my work begins. When I set out on my daily walk on such days, it's not for one of the carefree, meditative city rambles that I love. Rather, I keep my eyes to the ground and stop frequently to lift a stranded worm off the sidewalk and set it in a curbside garden or in the shade beneath a tree where it can retreat into the moist, cool, darkness. The earthworm is a "skin breather." In the absence of lungs, it absorbs oxygen directly into its blood vessels through its skin, kept moist by mucus-secreting cells. If the thin skin dries out, the worm suffocates. Some worms that I encounter are flat, brittle "S's." I can't save them. But some are dark, dry and barely moving. They're the ones that need me.

I rarely noticed earthworms, much less rescued them, until I read Charles Darwin's *The Formation of Vegetable Mould, Through the Action of Worms, With Observation of Their Habits*, a slim volume in which he makes no mention of evolution, natural selection, or the descent of man. By "vegetable mould" Darwin meant humus-rich topsoil or the A-horizon, the layer of soil in which I plant marigolds, beets, and snow peas. For forty-four years, Darwin closely observed earthworms and speculated about their significance. In the final paragraph of this book, published just six months before his death in 1882, he proclaimed, "It may be doubted whether there are many other creatures which have played so important a part in the history of the world, as have these lowly organized creatures."

Darwin based this lofty claim on what he observed in a pasture near his house in Kent. In 1842, a layer of chalk was spread over the field. Eventually, the chalk dissolved but pieces of flint remained. In 1871, Darwin dug seven inches into the soil before he reached the rock layer. He calculated that the earthworms had covered the flint with an average of 0.22 inches of fertile soil per year. Give the worms a full century and they'd cover the rocks with two feet of soil. Give them a couple of millennia and they'd bury a city. The Roman ruins in England were so well preserved, Darwin said, because earthworms had buried them.

What made these remarkable feats possible is that in the process of swallowing food—bacteria, fungi, nematodes, manure, decaying roots, rodents, bugs, and a tossed apple core—earthworms eat a lot of soil and so move tons of organic material from the surface into the soil. What they excrete are "castings," pellets of soil well mixed with organic matter that they deposit in mounds outside the mouths of their burrows. These castings are rich in calcium, nitrogen, potassium, phosphorus, organic minerals, and nutrients, and so fertilize my garden. And, too, tunneling earthworms break up hard-packed soil, sift and aerate it, even move stones. When in need of fish bait, my son goes in the backyard after dark with a flashlight and picks up the night crawlers that have surfaced into the cooler, damper night air. "Stay away from my garden," I tell him. "I need those worms."

Darwin didn't limit his studies to the earthworm's soil-building abilities. "As I was led to keep in my study during many months worms in pots filled with earth, I became interested in them, and wished to learn how far they acted consciously, and how much mental power they displayed." To test their sense of smell, Darwin breathed on them after he'd put tobacco, perfume, paraffin, and ascetic acid in his mouth. Worms only responded to the latter, Darwin surmised, because it irritated their sensitive skin. To test their sense of hearing, Darwin assaulted them with "the shrill notes from a metal whistle . . .

the deepest and loudest tones of a bassoon . . . the keys of a piano . . . played as loudly as possible," none of which had any observable effect on the worms. But when he placed a pot containing two worms on the piano and hit C in the bass clef, both worms "instantly retreated into their burrows." Earthworms couldn't hear sounds but they could feel them.

To study the earthworm's mental powers, Darwin observed them as they pulled leaves into the mouths of their burrows. One might think keeping food nearby to plug the entryway is an instinctive behavior. Yet the skillful manner in which these blind, deaf, and appendage-less creatures dragged the leaves, tip first, in most cases, and "foot stalks" extending from the burrow, and that their favorite leaf, that of the lime or linden tree, was not native, suggested to Darwin that they'd learned how and with what to stuff their burrows through trial and error. "A near approach to intelligence," Darwin concluded.

Sometimes before I drop a once-stranded earthworm into the shadier, damper soil beneath a tree, I pause. I observe the segments, stacked rings and grooves; the darker end where the miniscule brain and the voracious mouth are located; the clitellum, a smooth, light band that produces the egg case; the strong muscular contractions. If I had a magnifying glass, I could see the setae, bristles arranged in pairs around each segment that allow the worm to hang on tightly when a robin tries to pull it from its burrow. I marvel that this practically senseless, dim-witted, vulnerable, fluid-filled tube is both a fertilizer factory and an earthmover par excellence.

Recently, I discovered that the lowly but mighty earthworm is the subject of controversy. Now I'm not certain if it is my friend or foe. In 1995, Dave Shadis, a soil scientist for the Chippewa National Forest, noticed that the forest floor was changing rapidly near the shoreline of Leech Lake, a fishing resort area in northern Minnesota where my family vacationed when I was in middle school. Shadis noticed that duff, the thick, spongy mat on the forest floor created by decomposing leaves, bark, stems, and branches, was disappearing at a faster rate than it was being regenerated. As the duff disappeared, so too did the ground-nesting birds, woodland mice, and salamanders, who depend on it for food and cover, as well as ferns, wild ginger, and such spring ephemeral wildflowers as bloodroot, trillium, trout lily, and spring beauties. Shadis discovered that wherever he found evidence of earthworm activity, the duff was vanishing. Could there be a connection?

Native earthworms had worked the soil in Minnesota and the Great Lakes region until the most recent ice age, when the glacier scraped away the topsoil and the earthworms it contained. Consequently, northern hardwood forests developed in the absence of earthworms. About twelve thousand years later, when people from Europe and Western Asia arrived in North America, they dumped the ballast from their passenger and cargo ships—rocks, sand, and soil that they'd brought from their homelands—onto the shore. Mixed in with the debris were the seeds of purple loosestrife, leafy spurge, Canada thistle, numerous species of Eurasian beetles and earthworms, and other invasive, nonnative species. Since earthworms only migrate about a half-mile per century, they would have traveled only a few miles inland by now if left to their own devices. But with human assistance in the form of anglers who dump their bait in natural areas; the transporting of soil crawling with earthworms in the sand and gravel used in road construction; in the tire treads of off-road vehicles and lumber trucks; in the mulch; in the ball of dirt that accompanies trees, shrubs, and flowers for lawns and gardens; and in packages of earthworms for compost bins (one company sells a thousand worms for just $19.95!), the invasive annelids have spread throughout most of the United States. Humans are implicated in the invasion in yet another way. Peter Groffman, a microbial ecologist at the Institute of Ecosystem Studies in Millbrook, New York, says that it is climate change ("global worming" he calls it) that is luring invasive earthworm species into northern forests.

In the absence of earthworms, the loose soil of northern hardwood forests is covered with a thick layer of duff. This organic mat is vital to the forest since it regulates the temperature and moisture content of the soil, protects roots, nourishes and provides habitat for ground-dwelling creatures, and inhibits the germination of alien seeds. But because earthworms eat the duff faster than the forest can regenerate it, the once lush understory and diverse webs of species have been destroyed. Once the duff is gone, there are fewer sugar maple seedlings and an increase of species of invasive plants, such as garlic mustard and buckthorn. Moreover, as earthworms move the duff into the earth, they change the fungal-bacterial ratio of the soil. A more bacterial-dominated system hastens the conversion of leaf detritus to mineral compounds, thus robbing plants of the organic nutrients they require. Earthworm activity also makes the forest soils more compact by erasing the native mulch, which leads to greater soil erosion and the leaching of nutrients, which in turn degrades fish habitat.

My new knowledge about the actions of earthworms raises an ethical issue for me. What shall I do next spring when I find one of these ecosystem engineers expiring on the sidewalk after a big rain? Should I squish it beneath my rain boot or move it to safer ground? My answer comes from the ecologist Aldo Leopold, who famously said, "A thing is right when it tends to preserve the integrity, stability and beauty of the biotic community. It is wrong when it tends otherwise." My home in Lincoln, Nebraska, a city of more than a quarter of a million people on the far eastern edge of the Great Plains, was built on land that was once covered with lush, tallgrass prairie grasses and forbs. Here nonnative earthworms have been present for so long that an ecological balance appears to have been reached. That means that nonnative earthworms don't cause the destruction in this prairie place that they do in the forests rimming the Great Lakes. Here, where the soil is compacted by human traffic and cultivation, the earthworm is valuable because its burrowing activity brings a fresh supply of oxygen to the roots of trees, tomatoes, and tulips and its digestive processes break down nutrients, which makes them accessible to plants.

Next spring when I find earthworms drying out on the sidewalk following a heavy rain, I'll rescue them so they can aerate and increase the porosity of the gardens in my neighborhood, feed robins, provide good catfish bait, and continue to bear out Charles Darwin's final words about their remarkable influence: "It may be doubted whether there are many other animals which have played so important a part in the history of the world, as have these lowly organized creatures."

WORM HERDER
A Q and A with Dr. Diana H. Wall

CARRIE VISINTAINER

I took a coffee break with Dr. Wall in her office at Colorado State University to get the dirt on soil, worms, and the network of life beneath our feet.

Dr. Wall is the founding director of the School of Global Environmental Sustainability (SoGES) and a professor in the Department of Biology in the College of Natural Sciences at Colorado State University (CSU). Recently, she received the prestigious 2013 Tyler Prize for Environmental Achievement.

Let's cut to the chase. Do you love dirt?
I'm not sure that I love dirt. I *respect* it. I respect soil.

What are some of your early experiences with science?
Well, I grew up in North Carolina and urban Kentucky. My mother had been a biology teacher before she had kids, and she was the one who would say, "We're going camping" and encouraged me to join the Girl Scouts. This got me interested in the outdoors and biology. I was a Girl Scout until I finished high school because they had the coolest canoes, and they had access to the river.

When did you decide to study biology?
I studied biology in high school and I liked it. In college I took a microbiology course and found it fascinating. The professor was kind of old school and there was a test every Friday, and when you walked in the door you had to know everything. So it made you learn. And it made you look at all of these "small things" under the microscope, which was amazing.

How did you start working with worms?

Actually, I happened to be in the right place at the right time. In graduate school I was offered a research associate position in the department of plant pathology. I discovered that some of the small worm parasites of humans and animals can also be parasites of plants. They're very specialized. That's how I got into nematodes and roundworms.

What's the best way to look at nematodes?

These guys are the size of an eyelash, so you take soil and wash it through something similar to a kitchen sieve and what's left on the sieve you put into a dish under a microscope. There's so much diversity in a handful of soil, and this process was like entering a new world — seeing worms that had such identifiable features and lived in soil and could find their way to a certain plant root. And only on that plant root could they say, "I'm going to attack you." That's what made the difference for me. Farmers where I grew up already knew about nematodes, but I came in the back way.

What are some of the characteristics of different worms?

Since they're transparent, when you look through the microscope, you can see their esophagus — you can see them eating. You can see their whole reproductive system and determine a male or female. You can see a tooth inside them if they're a predator or a little spear inside them if they're going to attack a plant.

How many different types are out there?

We don't know. There are so few people working on this. Worldwide, there are only about twenty thousand species described, mostly in agriculture or forestry.

Why is it important to understand these worms?

Just like you see a food web above ground, there's a whole food web below ground that we depend on. All of these tiny animals and microbes are interacting to turn over nutrients for soil fertility. It's quite amazing.

How do they work together?
In all natural systems, they have a variety of jobs. They might be
feeding on plant roots, helping the plant in some way. There's also
bio-control, which is similar to predation—a balance of keeping the
population of bad guys down. One group eats only fungi, and others
bacteria. It's a network of decomposition, which stabilizes the level of
carbon in the soil and its release into the atmosphere.

What was one of your most important epiphanies during this time?
I would say the importance of these small creatures that we can't see,
that live in our soil and impact our well-being. It's similar to what we
thought about oceans years ago—they have diverse food webs that
we didn't know about, and we treated the oceans poorly. Well, we
have an ocean of soil beneath us that we're realizing is not a dump
either. It's providing us with our food. When we seal it over with concrete, we need to be a little careful about whether we're covering up
our most fertile ground.

Your work has taken you to Antarctica more than twenty times.
How did you begin your research there?
I had a post-doc at the University of California at Riverside, where my
main focus was to understand why there are so many species of nematodes and their role in the ecosystem. I was looking at nematodes in
the four desert ecosystems of the west, because I thought there would
be fewer species. It was crazy. There were so many! We "connected"
the nematodes to the plant or tree they were near. After fifteen years,
my colleague and I decided we needed to go somewhere where there
were no plants and plant roots, to see what, if anything, was there.
There are large "Mars-like" areas of soil in Antarctica.

What did you find?
In particular, we found one nematode, *Scottnema*, that was remarkable. It was first described by the British explorer Robert Falcon Scott
on a race to the South Pole. In the driest, most barren soils, we found
this nematode. It can survive cold, extreme weather changes, wind,
and manages to survive in soils with almost no organic matter.

What is Scottnema's *role in the ecosystem?*

It contributes significantly to carbon turnover—an important part of the decomposition process—in a way that is pretty incredible for a single species.

When did you begin to look specifically at climate change?

In the late '80s, we began to put down warming chambers in our research sites in Antarctica—huge circular chambers that slope off the ground, with an open top. The temperature inside was about two degrees centigrade (which is warm for Antarctica). Then we'd sample the soil each year. Our helicopter pilots named us "worm herders."

Worm herders?

Yeah, the pilots who flew us from town to our site would say, "What are you guys doing down there?" They thought we were herding worms into a cage and sampling them just like cattle. Which we were, actually. I thought it was a good analogy.

*What did you discover when you sampled
the soil from the heated chambers?*

I think the biggest thing that we've seen is well over a 65 percent decline in *Scottnema* over the years of our research, which is pretty darn fast.

Does this scare you?

Yes. Climate change is my biggest concern. We have to have nations act. But I feel optimistic. There are good things happening, locally and globally.

What's an example?

I'm very excited to be leading up an effort to compile a "Global Atlas of Soil Biodiversity." It's modeled after the "European Atlas of Soil Biodiversity," which came about because people are so worried about the loss of our soils, and because we can identify species and where our soils are vulnerable. I don't think this would have happened fifteen or twenty years ago. But now the science has advanced enough for us to translate the information for policy makers. It's become obvious that we need to protect our soils, just like we do our air and water. I'm excited to help frame the new science.

What will the Atlas look like?
There will be pages dedicated to each organism group, like microbes, and their job, and distribution maps. The book will be hardcover and large, so policy makers have to look at it in their office. They can't stuff it away out of sight.

When will it be finished?
We're working fast. It should be released by autumn 2015.

Who is funding this project?
csu and the European Union.

*Speaking of exciting things, congratulations on receiving the
2013 Tyler Prize for Environmental Achievement. What was your
reaction when you heard this news?*
I was stunned. I was totally stunned. I thought they dialed the wrong number.

You really had no idea?
No. It's the oldest environmental prize! Look at the scientists who have made breakthroughs and won—Keeling and E. O. Wilson, and so many more.

What will the prize mean for your research?
Well, science funding is always hard to come by. It will make a huge difference for my lab. There are lots of young people who want to learn more, make a difference, and think of solutions. This will support that.

*Another "award" of sorts is that Wall Valley, Antarctica,
was named after you. How did that come about?*
It was so funny. I had no idea about it. My colleague at Dartmouth College was searching on the Internet, and he happened upon this

information. This was three years after they named it! He called me up. We laughed. We decided we'd have to get a helicopter to take us out there in our next field season.

Did you go?
Yes.

So, is Wall Valley a prime piece of real estate?
Ah, not exactly. I don't think there's going to be anyone living there for a long time.

Maybe that's a good thing.
Yeah, it's pristine. A park.

Dr. Wall smiles and stands up from her chair. She shakes my hand. Her passion is contagious; she's the perfect spokesperson for the tiny animals that build and define our soil.

SEEING SOILS

*

DEBORAH KOONS GARCIA

Most people are soil blind. They walk on soil, they gaze at it on the horizon, they gain pleasure and sustenance from its bounty, but soil itself goes unseen, unappreciated. Modern life conspires to remove us from any connection to or awareness of soil. We spend a lot of time looking down, not at the soil, but at the various devices that connect us to our techno-fied world. Those of us who understand the importance of soil can bemoan the sad state of affairs as evidenced by the story of the schoolchild who visited a farm for the first time and, when presented with a carrot freshly pulled from the earth, proclaimed, "I'm not going to eat that—it's been in the ground!" We humans have a deep and essential relationship with soil but most of us don't know it. Because of this we treat soil like dirt.

Several years ago I decided to make a film about soil. I had made a film, *The Future of Food*, that was quite well received and I figured I would do a kind of follow-up, something in the same realm. I didn't realize until I entered the luxurious phase of research and settled down with my bookcase full of alarmingly thick soil textbooks how little I knew. In one of my early conversations with an esteemed soil scientist, she mentioned that she didn't see soil as an agricultural medium. I knew that she'd done decades of work focusing on agriculture so it struck me that unconsciously I was seeing soil as primarily an agricultural medium. I began to understand that soil is a vast and mysterious realm and agriculture is just one problematic part of that realm. Our relationship with soil goes back hundreds of thousands of years and agriculture has only been around for ten thousand of those. If our relationship with soil rests only on agriculture, our concern is too often about what we can get out of it, what's in it for us. That's not a very good foundation for a relationship.

As I went through the four-year process of researching, shooting, editing, and completing the film, I found that my attitude toward soil completely changed. At a certain stage, I didn't want to have any agriculture in the film. I empathized with the attitude of some Native American tribes—why would I cut into my mother the Earth? The plow, I'd learned, had caused more damage to this planet than the sword. I became protective of the soil and wanted to move away from the assumption that soil is a thing to exploit. I wanted to support and encourage a healthier relationship with what I came to see as a wondrous substance and bring that awareness to a wide audience.

The way I see soil now, if I were holding some soil in my hand and said, "This is soil," it would be like holding seawater in my hand and saying, "This is the ocean." While making the film, I wrestled with crystallizing and distilling complexity. The study of soil involves chemistry, physics, anthropology, geology, geography, but most importantly, biology—the study of life—because soil is the essence of life. Soil is about transformation. Without soil, there would be no human race. In fact, the word "human" comes from the same root as the word "humus," a component of soil. Soil is one of the true miracles of this planet.

So after many months of book learning and befriending soil scientists to find out what they knew and how they thought about soil, I started to shoot. I set up the first shot of soil, my cameraman framed and focused on a patch of ground, and I called out "action!" He turned to me and said quizzically, "There's nothing happening." "There is, there is!" I enthused. I realized I had to figure out how to take soil, which is dark and seemingly inert, and film, which is about light and movement, and bring them together in a satisfying work of art and science. How to take the highly technological medium of film and use it to connect people with nature? How do I lead my audience from being soil blind to being soil lovers?

My task as a filmmaker was to allow, invite, encourage, fascinate, and seduce my audience into understanding and feeling connection with soil. I observed during the editing process that when people who had no particular interest in soil would look at a sequence we were working on, they were mesmerized, completely fascinated by the soil orders or composting or mycorrhizal fungi. I figured that we humans have a deep and essential need to focus our attention on soil that goes back to our time as hunters and gatherers, when reading soil, plant varieties, and the sky and water around us was necessary for our survival, when observing land and changes that happen from day to day, month to month, year to year, drawing conclusions based on those observations, and taking action to use that knowledge and wisdom allowed us to survive and thrive on this planet Earth. We are by our very nature, in body and mind, connected to the soil. If I could wake up our human instinct to understand soil and its place in the world around us, I could help people nourish a part of themselves that gets starved in the modern world.

How to see soil: as an entity, as an organism, as an ecosystem, as a community, as a collection of cycling processes and nutrients, as a place where billions of microorganisms interact, as a plant-growing medium, as a living system. It's alive! We can see soil in time and in space. We see soil in close-up, medium, and long shot. Most importantly, I decided to present soil as a protagonist of our planetary story and the protagonist of my film. Soil is born, it has parents, a life cycle, different characteristics. It is the zone of transformation on this planet. Soil is busy being born and busy dying all the time.

I decided to present the story of soil in a classic three-act structure. Act 1: Soil itself, its multidimensional nature. Act 2: Soil in relationship, primarily in relationship with humans. Inevitably, that relationship must focus on agriculture. Act 3: Soil and big-picture ideas such as soil and global warming, soil and water, soil and feeding the world, soil and metaphysics.

Having digested so much material and having debated with myself and my colleagues about what should be in the film, I came up with a wonderful title: *Symphony of the Soil*. The piece would be complex, made up of many different instruments and parts, creating a satisfying whole. Soil is a symphony of elements and processes, and my film would evoke that. This also allowed me to bring powerful music into the mix, to sweeten the science.

To further my goal of promoting soil consciousness, I decided to embrace science, art, and activism, to create a hybrid that is all too rare in film these days. I found a young artist who painted hundreds of beautiful watercolors to animate ideas like photosynthesis and the nitrogen cycle. This tactic adds variety and texture as well as understanding to the film. In all the decisions I made, I committed to a process in which meticulous science is honored. Filmmakers these days are often wary of presenting too much science, and they dumb things down. I am all for a smarten-up strategy. I believe people find science deeply satisfying, if it's presented in an intriguing way. The word science comes from the Latin word *sciere*, to know—science is about what we know. People want to know, they are hungry for information that helps them make sense of the world.

Soil science is a field where much is known and much more remains to be discovered. Soil science is now cutting-edge science. For example, soil can sequester carbon, and much research is being done to figure out how treating soil right can help ameliorate climate change. Soil scientists appreciate what they do not know yet, and this makes them quite appealing. They understand mystery and this raises their work above the mundane. We can feel and share their enthusiasm. The study of soil provides fertile ground for exploration and active engagement. In order to fully appreciate soil, you have to understand it.

Soil is a transformational substance. It's the place on earth that transforms life into death, death back into life. In the film, scientists dig soil pits, then stand in them to explain what's happening in that particular soil profile. During the first few screenings for lay audiences, I was surprised by the laughter that occurred when the soil pit shots came up on the screen. I realized that there is a subconscious connection between a soil pit and a grave. The laughter was a reaction to that. Examining a soil profile in a curious way makes us face our own mortality. Coming face to face with a soil profile reminds us that we arise from the soil and that we return to it.

Because of technological advancements such as electron microscopes, satellites, and time-lapse photography, we are able to observe soil from more vantage points than ever before in history. Soil in extreme close-up: We can see soil microscopically and start to learn about the billions of microorganisms that are at home in a healthy soil. We are just beginning to understand the relationships between these microorganisms and their dynamics within the soil. On the most microscopic level, ions are constantly moving between elements.

Soil on a human scale: We can experience soil by using our eyes, walking on it, smelling it, feeling it in our hands, observing the kind of plants and creatures that live on and in the soil around us.

Soil on a planetary scale: We can see soil in long shots, from satellites, from space, and track the changes that climate, the environment, and our human practices make in soil. We can also see soil in time, using technology to understand what happened in the past and how that affects what's happening now and could happen in the future.

One day as I was in the middle of making the film, I went for a walk in the hills I've been exploring for forty years and felt fully conscious of how much movement, how much activity was happening in my own soil community. From mountain lions, deer, foxes, voles, worms, dragonflies, dung beetles, bacteria, fungi, springtails, all the way down to electrical charges bouncing around—it was all about movement. Awareness of the processes and cycles and organisms of all sizes that were in play, all at that moment affecting my world yet unseen, invisible to me, expanded my vision exponentially.

One of the interesting things I learned about soil is that it's not just "earth." Soil is 25 percent air and 25 percent water, and that water is held in the soil, no matter how dry it seems. The other 50 percent is solid matter, mostly made up of minerals. Some of the solid material is the famous SOM, soil organic matter, constituting from 1.5 percent to 15 percent of soil mass. Most of that SOM is made up of dead microorganisms, and part of it, a very small percentage, is alive. I see soil as being made up of all the elements—earth, air, water, and the microorganisms that are the fire, the force that drives life as it cycles through the soil.

We now know that soil is a complex ecosystem and a mysterious one. Ninety percent of the microorganisms in soil have not been identified, much less understood as to how they function. These microorganisms form communities that, along with infinite numbers of other organisms, make up the soil community, and we humans are a vital part of the soil community. From the smallest microorganisms to insects to prairie dogs to owls to bison to humans, we are all taking from and giving back to the soil. Soil is the matrix and web of life, and it's all about relationships. Burrowing animals churn up the soil to let air and water in, to let roots grow. Without them, soil can become compact and dried out. Without grazing animals, prairies would become forests. Prairies support grasses, and when their deep roots die they replenish the soils, which is why grasslands tend to have the highest fertility of all soils. By contrast, tropical soils are shallow and relatively infertile because all the life is held in the dense forest plants. We humans have not always been particularly good members of our communities. When a leaf falls from a tree and decomposes back into the soil under that tree, it "gives back" to that tree the nutrients it took from it. Modern agriculture takes from the soil and, too often, does not give back, depleting soils and causing long-term degradation. Practices such as composting can give back, mimicking the cycle of nature.

At one point during my fascinatingly soil-centric life, I was invited to go to China to show my films. One of our hosts was an honored scientist who told me that only about 12 percent of China's soil is usable for farming or grazing. By contrast more than 43 percent of U.S. soil is really good, either mollisols (prairie soils) or alfisols (forest soils). This thoughtful Chinese academic told me that the reality of so little really good soil has shaped Chinese culture and character: they are more collective, more careful, more conservative because they have to be. They have to pull together to be able to feed themselves. They have a similar situation with their water.

The American character has also been shaped by our natural resources; we have lived as if we are so rich in land that we can abuse our soil without consequence. We think we can poison our fields, let topsoil erode and blow away, and there will always be more, somehow, somewhere. We've already lost and destroyed an alarming amount of our fertile soil. With so much abundance, we believe we have little need of being careful and husbanding our resources. We don't like limits, because we imagine we don't have any. China has been around for thousands of years, our modern U.S. society about three hundred. Our soils have become profoundly stressed and degraded.

If we keep farming the way we are now, we could be out of topsoil in thirty years. The sad fact is we are facing peak soil and peak oil. And there's a connection: our oily industrial agriculture is not kind to soil. We need to reconsider what we assume is our national character, our "natural" tendencies. We've taken much from our soils and not given back. In our society now, taking, taking, taking and not giving back has become admirable in some circles. This does not create a healthy community or society. As a supposedly highly individualistic culture, we feel limited if we are enjoined to think of our communities before our own gratification. We don't like limits, but we are running up against them. Rather than slipping into arrogant denial, we need to learn to thrive within limits and to give back to create a more wholesome life for everybody. We must take that as our nation's life work. It's time to draw upon an entirely different aspect of our character: we are up against many profoundly difficult challenges and our ingenuity and industriousness can be employed to meet them. Magical thinking won't help us or our soils. Understanding and right action will.

Soil lovers unite! The planet we save will be our own!

THE NEXT BIG THING
IN SOIL SCIENCE

✻

CARL ROSEN

*S*oil scientists and soil microbiologists in particular have known for a long time that soils in natural and managed ecosystems are quite complex. For example, it is estimated that one gram (about one teaspoon) of soil from a garden or a natural prairie may contain one billion bacteria and more than five thousand species of bacteria. This means that the population of bacteria in three tablespoons of soil is more than the population of all the people on earth. And this is only bacteria; there are other organisms in soil including fungi, algae, nematodes, earthworms, insects, and spiders, as well as numerous burrowing animals. One of the grand challenges in soil science is to determine how much biodiversity there is in soils and what role these organisms play to sustain life on Earth.

Although I grew up in a suburb of Philadelphia with little appreciation for the importance of soils, it was an opportunity to work on a dairy and potato farm in Ireland that sparked my interest in learning more about how our food is produced. Before I left that farm, the farmer I worked for talked about the importance of giving back to the soil and the use of manure from his dairy cows for that purpose. This experience led to my college studies in and passion for the field of soil fertility and plant nutrition. While fertilizer use has helped feed the world population over the past century, more efficient use of nutrients is needed to prevent unintended consequences such as degradation of surface and ground waters. Nutrient cycling is mediated to a large extent by soil biology. Therefore, a better understanding of the factors that control the presence of and interactions among organisms in the soil will provide us with ways to more sustainably manage this natural resource that humans often take for granted.

Soils are basically composed of three fractions: air, water, and solids. The solid fraction is composed of organic matter and the inorganic particles — sand, silt, and clay. Water contains dissolved nutrients essential for plant and animal life. Soil air contains, among other gases, oxygen needed for respiration by plant roots and microorganisms. The ability of a soil to function as a medium for plant growth is largely controlled by interaction of its biological, physical, and chemical components — the balance between water and air and the availability of nutrients. With too much or too little of any of the three fractions, little will grow, and that soil gets called dirt.

Soil organisms have a strong influence on both the physical and chemical properties of soils. They use carbon as well as other plant and animal components in residues as a food source. In the process they form humus, which in turn stores carbon and interacts with the inorganic particles of soil to improve soil structure. The ability of soils to drain or hold water is in part a function of soil structure. Even though, on average, in agricultural and natural systems the organic matter makes up less than 5 percent of a soil by weight, it allows the soil to provide the conditions necessary for life above the soil to exist.

So with the complex biodiversity in soil, what are the ways we can learn about how soil microorganisms interact and perform their important functions? Measuring soil respiration and determining different carbon and nitrogen fractions has been used to help quantify soil health, but it does not identify which microorganisms are involved. One approach is to study soil in harsh environments that limit biodiversity yet allow scientists to study how those organisms respond to the harsh environment and survive or interact with their surroundings. While that approach has provided interesting results, different and novel approaches are needed, because only a small fraction of the microbes in most soil ecosystems can be cultured in the laboratory.

A recent advance in documenting and understanding soil biodiversity involves genetic testing. Similar to learning more about the genetic makeup of the human genome using molecular biology tools, soil microbiologists are now extracting the DNA from soil organisms in an attempt to understand the biological complexity of that environment. Genomics, or more precisely metagenomics, is being used to sequence the DNA from soil microbes to try to identify and understand how soil microbes might actually function in the deep dark below us. But adding metagenomics to the mix has the potential to yield great results. In the lab, we can extract the DNA from soils managed in different ways—for example, an organically managed system versus a conventionally managed system. Practical results may take a long while to achieve, but a courageous plunge into the varied soils beneath our feet calls to all of us in the soil sciences.

It is the next big thing in soil science. We are just getting started, and I am excited to be here at its inception.

A BADGE OF HONOR

*

TOM WESSELS

*M*y first real connection with dirt happened as a boy making a baseball diamond. Andy's back yard was bigger than others in our neighborhood and had two sections. Right behind his house was a classic lawn separated from the undeveloped, farther portion by a small, slanting bedrock ledge that ran down to a forty-foot-wide terrace. Bordered by what we kids simply called the Woods on one side and the ledge on the other, and covered in recently exposed, raw dirt from the removal of stumps by bulldozer, the terrace was the perfect spot to create a ball field for young boys.

My job was to help lay out the baserunning paths and remove all rocks from them. After that we seeded the infield and outfield, built a large backstop out of scattered, waste lumber scavenged from our various homes, and by midsummer started playing ball.

I loved sliding into bases on that diamond—even when I didn't have to—just because I had prepared the rock-free paths and it felt so cool. My mom, after weeks of trying to tell me to keep my jeans clean, eventually acquiesced and a couple of pairs were relegated to playing ball. To this day dirt smeared into my pants feels like a badge of honor, but one I never truly appreciated until a few years ago.

I'm a terrestrial ecologist and love roaming the woods and mountains that grace my native New England. In the woodlot of our Vermont home I developed a one-mile loop path that is free of everything but pine needles. I have walked that path in the dark of night and can find my way solely by the feel of the ground beneath my feet. Since the path is wide enough for my pickup to bring in firewood, it has been compressed over the years. One step off the path and I quickly feel the springiness of forest soils that have developed for more than a century since the former dairy pasture was abandoned to grow back to forest.

The dirt in this woodland is quite different from that of my boyhood ball field. It is spongy, the accumulation of more than a century's worth of humus. It is also loaded with life.

As an ecologist I love the rich diversity of species found in New England, from wild flowers and trees to birds, mammals, and bees. For the majority of my professional life I'd never considered the diversity of life in the dirt on which I tread while marveling at the forest flora and fauna.

Then we hired Rachel Thiet to join the faculty of the Department of Environmental Studies at Antioch University New England. As a soil ecologist Rachel told me one day to be careful about my "above-ground bias." By this she instructed me that the biotic diversity below ground dwarfed that above ground and in fact all the above-ground species were completely dependent on the biotic community residing in dirt for their existence.

For the past decade I have taken Rachel's teaching to heart. Now when I step on the soft soil of a one-hundred-year-old forest or the deep, spongy soil of old growth, I sense the thriving diversity under my feet. I have also learned some amazing things that lay hidden in dirt—particularly regarding mycorrhizae.

Mycorrhizae are one of the four major groups of fungi in terms of their ecological roles. The other three include decomposing, parasitic, and lichenizing fungi. The role of mycorrhizae is to associate with the roots of photosynthetic plants, where they extract carbohydrates as their source of energy. In return, due to their extensive mycelium throughout the soil, they allow plants that interact with them to increase nutrient uptake manyfold. The majority of our coniferous tree species, as well as orchids and heaths, have to interact with mycorrhizae in order to survive. So plants and these fungal partners have developed a broad network of mutually beneficial interactions.

Mycorrhizae are quite egalitarian in that one fungus can interact with many different plant species at the same time, thus linking all those plants together. In 1997, a mycorrhizal study done in a forest of Douglas fir and paper birch came up with an astounding finding —one that dramatically changed how forest ecologists view these ecosystems.

The intent of the study was to calculate how much of a tree's energy was paid out to its mycorrhizal partner. To do this, Douglas fir were treated with carbon dioxide that was composed of the rare isotope carbon-14, to be used as a tracer. Knowing this carbon dioxide would be incorporated into carbohydrates manufactured by the trees during photosynthesis, researchers could follow how much of the carbon-14-based carbohydrate ended up in the mycorrhizae. Through this protocol they figured out that the Douglas fir paid out about 15 percent of its energy to its fungal partner, not a bad deal if you need that partner for your survival.

Then one of the members of the research team decided to examine the paper birch in the forest that had never been treated with carbon dioxide containing the rare isotope. Up to 6 percent of the energy in the paper birch was carbohydrates composed of carbon-14 that had been manufactured by the surrounding Douglas fir. This was the first study to show that mycorrhizae not only transferred nutrients but also energy, even between different species of plants. More recent studies now show that mycorrhizae can also channel information between plants, making them an ancient and highly effective precursor of our vaunted Internet, all hidden in dirt! We used to think that trees simply competed with each other for forest resources. Now we see that it is far more complex than that, with the mycorrhizae weaving a resilient network where trees actually support each other.

In a way, we are in our infancy in understanding the incredibly complex life of soils. Recently in one of my graduate courses a student told the class of an experience she had in her undergraduate microbiology course. She and her lab partner were examining the bacterial community in a sample of compost. By culturing the bacteria and then doing a DNA analysis they identified a completely unknown species. I have heard of high school students doing soil studies and identifying new species of invertebrates. Dirt may be the least understood ecosystem on our planet!

I have also learned to appreciate the stuff based on how it affects social systems. The nature of dirt very much molds the culture of the people interacting with it, often without their even understanding this relationship. Obviously moist and nutrient-rich soils will foster the cultivation of crops whereas drier ones may support grazing, giving rise to different farming cultures. But dirt's influence can go even further, impacting entire political economies.

My favorite example of this involves the sister states of New Hampshire and Vermont. Although the two states lie side by side at the same latitude with the same climate, their political economies differ. This difference is directly related to the nature of each state's dirt.

New Hampshire is underlain with predominantly drier and less fertile soils derived from weathered granite, while Vermont has moister, more nutrient-enriched soils generated from its calcium-laden metamorphic rock. Due to their respective soils, the early economy of Vermont was very much based on farming. In each town across the state, similar-sized farms were the backbone of its economic system. Therefore no one landowner or town had more economic clout than any other. So Vermont's early political system was one where power was equally shared.

In New Hampshire, where the dirt didn't favor agriculture nearly as well but produced valuable oak and pine trees, the early economy was based on timber. With this came timber barons who controlled large portions of this economic sector, giving them greater political power. With this difference the two states' political systems moved apart. The New Hampshire system favored large business. In Vermont it was more closely aligned with the citizenry. These tendencies continue to this day and can be clearly seen in differing state policies.

Vermont was one of the first states to pass a bottle return bill, pushed by dairy farmers who feared their cows' feet would be cut by bottles cast into pastures from passing vehicles. New Hampshire is the only northeastern state never to pass a bottle bill—the rationale being it would be too costly for business. Billboards are banned in Vermont because its citizens don't want their rural vistas impacted by advertising. Not so in New Hampshire. Vermont has rigorous land development standards to guard against sprawl or growth patterns that would impact the state's rural character. New Hampshire's policies are friendlier to development. Who might have guessed that this is all due to dirt!

I have former students who work small farms that produce organic vegetables by working closely with the earth. I am sure they would not be surprised by this sister-state analysis. They clearly understand the central role dirt plays as the lifeblood of their farms and lives. Why shouldn't it also impact even larger things like state policies? As an ecologist, I'm still surprised it took me so long to understand this. Like so many others, I missed the importance of what resides hidden below my feet. Now as I move through the latter portion of my life, I am encouraged that I am still learning big things. Dirt—the stuff that is often regarded as an enemy to cleanliness, or something to be scraped away by bulldozers—is in fact big. It is essential for the well-being of everything in our regional landscapes and because of that our personal welfare too. Now when I rise from tending our vegetable garden and see my dirt-smudged pants, I am gratefully aware of being able to wear such a noble badge of honor.

DIRTY BUSINESS

*

DAVID R. MONTGOMERY

I came to love dirt from below, through studying the rocks from which it comes. As a fledgling geologist, I learned to read Earth's autobiography by looking past the soil. My training emphasized how the world's blanket of weathered rock and organic matter obscured the view back into deep time. Soil was not the interesting stuff, or so I was told.

But I was never one to believe all I was taught. Not long after I declared a geology major, I found a copy of *Topsoil and Civilization* in the bargain bin at the Stanford bookstore. This dusty, long-out-of-print gem opened my eyes to seeing soil as the foundation for human societies. As I gravitated toward studying how landscapes evolve, I came to appreciate the importance of living soil for humanity's future. Working around the world, I saw how the history of humanity is written in dirt and how the fate of civilization likely depends on how we inscribe the next chapter on the land.

I first came to appreciate the way nature can reclaim a city, and begin turning it back into soil, in the Guatemalan jungle at the ruined city of Tikal. There, wandering off the main path, I came to a forested hill with a footpath leading up. Anticipating a fine vista overlooking the ruins, I climbed, passing pried-up blocks of gray limestone intertwined with the roots of large trees filtering light to the jungle floor. Partway up the hill, I noticed that some of the blocks of rock were inscribed with carved Mayan symbols. At the crest of the hill, remnants of stairs led to the ruins of a small chamber perched right above the jungle canopy. It dawned on me that the whole hill was a building, a massive temple reclaimed by the forest. Then I looked around and realized that all the hills I could see were really ruined buildings. In less than a thousand years, the jungle had reclaimed these man-made hills, covering them with a fresh blanket of soil. I would never forget how, given enough time, nature could reclaim a city and cover it with fertile soil.

A decade later, driving through the lower Amazon in Brazil, I saw how fast human actions can strip a landscape of its soil—and its ability to sustain life. After completing fieldwork in a remote, still-forested region, we bounced along a rough dirt road heading back to the first major town. The first villages we came to were collections of subsistence farms where the forest had just been cleared, the red-brown soil freshly exposed and planted. A dozen or so miles farther from the forest, and closer to town, the soil was gone. Yellowish, rotten rock lay exposed at the surface of fields with sickly crops. A few more miles toward town, the villages were replaced by grazing cows that could survive on the sparse grass that grew on the degraded land.

Here was the untold story behind rain forest beef. Subsistence farmers desperate to feed their families cleared the forest and planted crops, but they caused enough soil erosion that the soil disappeared in less than a decade, forcing the farmers deeper into the jungle to clear new land. Landholders running large herds of cows moved in only after the farmland was abandoned. Seeing firsthand how these desperate farmers were putting themselves out of business sent me off to explore parallels to the experience of societies around the world.

I discovered that civilizations from Babylon to Easter Island proved only as durable as their land. Archaeological studies in Greece, South Pacific islands, Central America, and other regions implicate soil erosion and degradation in the decline of ancient societies. Although the reasons behind the rise and fall of particular civilizations are complex, time and again the state of the soil set the stage upon which economics, climate extremes, and warfare assured their fall.

In a broad sense, the stories of many civilizations follow a common story line in which agriculture in fertile valley bottoms allowed populations to grow to the point where they came to rely on farming hillsides, where tilling exposed bare earth to rapid erosion from rainfall, runoff, and wind. Throughout history, societies grew and prospered as long as the soil remained productive or there was new land to plow, and declined when neither remained true. The trigger for societal collapse may have been a drought, natural disaster, or social conflict, but the resilience of societies lay in the state of the land, in the health of their soil. In both small, isolated island societies and extensive empires, soil erosion and land degradation limited the longevity of civilizations that failed to safeguard the foundation of their prosperity —fertile soil.

History records many examples of how, under the right circumstances, climatic extremes, political turmoil, or resource abuse can bring down a society. It should be sobering that in the upcoming century we face the potential convergence of all three, as shifting climate patterns and rising oil prices collide with accelerated soil erosion and ongoing degradation of soil fertility. While the first two issues are widely recognized, the twin problems of global soil loss and degradation receive far less attention.

Soil degradation remains one of humanity's most insidious and underacknowledged challenges. We have already degraded nearly a third of the world's crop land, much of it in the past half-century. China just admitted to seriously polluting a fifth of its farmland in recent decades. In the United States, we continue to lose farmland at a rate perhaps too slow to notice year to year, but alarming nonetheless when one ponders how to feed the world's growing population later this century. In 1990, the Global Assessment of Soil Degradation found that human-induced soil erosion and salinization had already affected almost two billion hectares of agricultural land. A 2008 report, developed by 400 scientists from 110 countries, concluded that business as usual is not an option when it comes to soil, food, and people. When I compiled data from around the world, I found that soil erosion under conventional agriculture generally exceeds rates of soil formation by more than tenfold. We are feeding ourselves by mining our natural endowment of fertile soil, by skinning the Earth.

In the coming century, we face the fundamental challenge of simultaneously adapting our agricultural systems to feed a growing population and safeguarding both soil fertility and the soil itself. Although the experiences of past societies provide ample historical basis for concern about the long-term prospects for soil conservation, data from recent studies indicate that no-till farming and conservation agriculture could reduce erosion to levels close to soil production rates. Similarly, organic farming methods have been shown capable of improving soil fertility; agricultural production need not necessarily come at the expense of the soil. In other words, we need not repeat the experience of past societies, even if we appear on track to do so.

Years after being advised to overlook the soil covering up the world's rocky bones, I have come to see that if modern civilization continues the boom-and-bust cycle of land degradation, we will ensure a global disaster when the agricultural math of a growing population collides with the geographical reality of a shrinking supply of fertile farmland. I've also come to see that we could rebuild soil fertility even while using land intensively, should we choose to. Others have done so in the past. Centuries ago, the Dutch reclaimed land from the sea by adding organic wastes back to enrich the soil. Long before then, Amazonian Indians and the Inca in Peru improved their soils by returning organic matter to the land. Likewise, the practice of returning night soil (human waste) to the fields built soil fertility on intensively farmed land in Asia. The common element? Returning organic matter to the land.

Agriculture has experienced several revolutions in historical times, and much like mechanization did a century ago, changes in farming practices could once again transform agriculture through widespread adoption of no-till and organic methods. The argument advanced by advocates of "conventional" farming that organic agriculture cannot possibly feed the world and would lead us down the path to starvation, has been challenged by studies showing that organic farming can produce both crop yields and profits comparable to conventional methods. Indeed, the highest per hectare crop yields generally come from small-scale, labor-intensive organic farms. A 2007 study from the University of Michigan analyzed a global dataset of 293 examples of conventional and organic crop yields and showed that organic agriculture could feed the world without expanding the agricultural land base. What would it take to do this? Restoration of native soil fertility.

Soils are incredibly complex, reflecting the influences of climate, biology, and the underlying geology. Despite this complexity, two simple principles make sense for guiding how to sustain soil fertility over the long run: don't let erosion race ahead of soil formation (don't run out of good dirt), and feed the life in the soil that drives nutrient turnover and recycling (sustain soil fertility). The key to sustainable farming is the idea that things favoring healthy soil life promote the health of life above ground. Those who insist that high-input agriculture is the only way to feed the world must face the simple truth that crops grow best in fertile, life-filled soil.

Here lies the basis for a soil ethic—a dirt ethic, if you will. Agricultural practices that build soil life and soil fertility are good. Those that don't are bad.

Adopting such thinking would usher in a new agricultural revolution, one we have a long way to go to implement. During the twentieth century, both the Haber-Bosch process—which produces about half a billion tons of nitrogen fertilizer a year—and the ambitious worldwide Green Revolution effort to develop fertilizer-intensive crop varieties, encouraged thinking that divorced agriculture from stewardship of soil life. Increased yields were propped up with intensive fertilizer inputs that had devastating effects on soil life. Mycorrhizal fungi, now known to play a fundamental role in delivering soil nutrients to nearly all plant species, experienced precipitous declines under chemical-intensive agriculture.

When chemical fertilizers were first used they were hailed for good reason—they boosted crop yields on degraded land. Thereby masking the effects of soil loss and declining soil fertility, the agrochemical recipe at the heart of modern agriculture obscured the ecological basis of soil fertility. A recent report sponsored by the United Nations and the World Bank concluded that industrial agriculture based on high external inputs is neither sustainable nor resilient. Unfortunately, fertilizer-intensive agriculture is as addictive as heroin. Once hooked, it's hard to quit.

Poisoning the foundation of our food web is a short-sighted strategy for sustaining life at the top (us). Yet this is the basic strategy behind relying on intensive use of pesticides and herbicides that degrade the life of the soil. Simple prudence dictates that over the coming decades we begin restoring life to the world's soils and rebuilding native soil fertility. This is not to say we must eliminate agrochemicals altogether, but that we must become wiser and more sparing in their application. We need to prioritize transforming conventional agriculture to restore soil fertility no matter what technologies are pursued to increase future crop yields.

And yet despite all the technical challenges, I think the hardest part about revolutionizing our agricultural system will be learning to see soil differently. But I know this can be done. Over the course of my career, I've changed how I see soil.

And in the past decade, I've been surprised to learn how quickly soil can be restored. Nature, after all, makes soil slowly. Yet my gardener wife restored our urbanized soil in north Seattle in less than a decade, transforming a barren yard with a ratty lawn into a vibrant urban jungle with several inches of new topsoil. And I've visited farms where aggressive measures to restore soil fertility have paid off with higher yields and improved profits within the span of a few short years. The key to restoring soil is to see it not as a substrate in which to grow plants, but as an ecological system for feeding them. We need to feed soil life so it can help feed the plants that feed us. What does soil life eat? Organic matter. Feed the soil and it will feed you (and the plants you eat).

Can we really restore soils globally? It would be quite a challenge, requiring changes in waste processing, agricultural policies, and farming practices. But if soil restoration can change urban soils into productive farms — as has happened in Detroit, Seattle, and the Bronx — then soil restoration can happen in the countryside too.

At the same time, we must consider the differing needs of developed and developing nations. In the developed world, the challenge lies in transforming large factory farms into technological hybrids that incorporate aspects of organic methods based on principles of agro-ecology. In the developing world, we face the challenge of how to feed the destitute too poor to buy food. The answer to this problem does not lie in proprietary crops and agro-technology for the simple reason that those with no money cannot afford to buy into those technologies. The key to feeding the world's hungry is to provide access to land where they can use their singular asset — their labor — to feed themselves. Small-scale organic farms offer a time-tested way to do this.

Fortunately, promising approaches are beginning to frame potential solutions to the problem of soil degradation. Soil-building methods can restore soil fertility remarkably quickly, given an adequate supply of organic matter. What do all these efforts have in common? They are all based on building soils while using them, by making soil fertility a consequence and not a casualty of agricultural production.

For all the attention focused on climate change, the end of the oil era, and loss of biodiversity, there is a real danger that society may neglect the most basic environmental change sweeping the planet. Even though it is hard to notice in a single lifetime, Earth's continents are losing their fertile soils in a process that, if sustained, will eventually undermine civilization. Yet soil restoration can also contribute to addressing climate change. Over the coming decades, changes in agricultural practices could significantly help offset global carbon emissions. Restoring healthy soils can also help produce better food and cleaner environments and can improve public health.

What stands in the way of soil restoration? For starters, soil has an image problem. Every nation's most strategic resource is the thin layer of rotten rock, dead plants and animals, and living microorganisms that blanket the planet. Yet we treat this natural capital like dirt instead of the frontier between the dead world of geology and the living world of biology, the living foundation for life on land. Our perspective on this really matters, because how we treat the land will, in turn, shape how it treats us. For in the end, life *is* a dirty business.

FEED YOUR SOIL

*

BOB CANNARD & FRED CLINE

*S*oil is life. The soil health feeds the plant, which feeds the man. It takes no more effort to treat the soil as the living entity it is, to nurture and enhance its health and complexity, than it does to simply use it as a means to maximize crops. Quick farming fixes tend to create quick disasters. Instead, we need to grow our soils as we grow ourselves: sturdy and strong, for long-term health.

I, Bob Cannard, am the director of Green String Farm, a 350-acre working farm in Petaluma, California. Green String is owned by Fred and Nancy Cline, who also own two wineries and many hundreds of acres of vineyards throughout northern California. All these properties use the Green String method, outlined below, and the results speak for themselves. Our produce is the first choice for Alice Waters, owner of Chez Panisse, one of California's premiere dining venues. Many top-shelf chefs, and locals stocking their kitchens, also choose Green String. In addition, Cline Cellars and Jacuzzi Family Vineyards produce wines of very high renown, and the estate and single vineyard varietals are all grown the Green String way. The Clines and I are passionate stewards of the land in our care, and we intend to see it thrive for many generations to come.

As with all life, soil life can grow or decline in step with the site-specific environment and the environment at large. It is our job as farmers to increase soil life and grow food in a nutrient-rich environment. Rooted in healthy soil, our food will nurture us and move us toward environmental completeness. It is our contention that if humanity were to focus on soil life and overcome deficiencies in the fields, providing complete nutrition in our food crops, then real, tangible human results could be achieved.

We do not view nature as an adversary, the grower versus various "obstacles" that must be overcome by chemical or forceful means for a successful crop. We hold to a different perspective, one that lends support to the natural process. This attention to and respect for what is underlies our fieldwork. It leads to the economy of plenitude found in nature. One bean produces hundreds of beans. One apple seed makes a tree producing thousands of apples. What could we achieve if humanity were to embrace this economy of plenitude? Widespread health and reduction of illness? A stronger, brighter, more balanced population? Climate stabilization? Increased employment? Tranquility? We could create a truly satisfied human culture. We could be satisfied in life.

SOIL IMPROVEMENT AND COMPLETE FOOD

Like humans, plants digest in order to grow. When a plant has good digestion it can absorb minerals from the soil. If the soil contains all the compounds the plant requires to form and function, it is satisfied and has no hungers. The plant develops to its full potential and imparts etheric completeness as well as mineral-based organic complexity. We work to create complete soils and to bring the nature of those soils to the minds and bodies of those who consume our food. Happy plants make happy humans.

Sadly, the life force in most tended soils has been diminished from misuse. But humans can work together as stewards of the land to renew the energy of the soils for current and future generations. We don't have to continue the "fast food" approach of agribusiness: dump on additives and reap hyped-up harvests. We can slow down and really feed our soil. We can engage in a conversation with the soil rather than making unsustainable commands. Learn to give as well as take. Listen and appreciate and respond to dirt's miraculous capacity for creation.

At Green String, we attend to four primary physical food groups in order to reinvigorate the Earth's soils. Yes, we feed the soil so the soil can and does feed us well.

1. DIGESTIVE CIVILIZATIONS

Healthy soil requires an uncountable multitude of microorganisms. These create a kind of civilization, achieving a balance of earthworms, viruses, bacteria, and arthropods. Each of these creatures lives and dies, creating a healthy ebb and flow. Though soils differ in the types of life contained, all soils have a common, universal need for this microcosmic civilization, based on digestive relationships. There is a very real urgency for people to use intellect, common sense, and an authentic emotional connection to increase and protect soil biology.

Diversity of soil civilization leads to soil health. A healthy soil is content; a weak soil strives for increased life. Weeds, nonnatives, choked growth—all are indicators of weakened soils. As stewards of the land, we must recognize and aid site-specific soil digestion. Get to know and honor those civilized neighbors beneath your feet!

2. A CRUSH ON ROCKS

The origin of soil is magma. Cooled and eroded by the eons into a home for microorganisms that digest it, rocks create the canvas on which other life can flourish. Varied, complex associations between different elements form the catalysts for most enzymatic systems. These interlocking systems provide the opportunity of life. The most actively available forms of soil minerals are those that have been heated and mixed in the magma and cooled quickly, as in ash, cinder, and lava. This mineral nutrition then benefits soil life forms, the plants that grow in the soil, and the animals that consume the resulting food. Complete mineral nutrition creates the opportunity for organisms to development to their full potential. Ask any tomato plant lucky enough to land in a lush volcanic valley.

In nature, soil is renewed and restored through volcanism. The atmosphere fills with ash, which settles on the soil and replenishes the eroded sediment deposits on the flood plains. In addition, you can follow water's course as it carries minerals from the mountains to the ocean. Snow and rain fall on the boulders in the mountains, and water breaks them down and carries the minerals into rivers and creeks. These minerals get into the soil and feed the plants that root into it. At the seashore, only sandy silica remains. Spent of its nutritive value, the dirt has done its work. We dam rivers and impede this vast flow of minerals into the soil at our own peril.

But humanity has the mechanical capacity to mill and distribute minerals to soils in need of replenishment. This is how we can easily overcome environmental nutritional limitations. And this broad-spectrum distribution of minerals will surely increase the life force of the dirt and the plants that root into it. The cost of such a system is small compared to the micromanagement associated with conventional chemical agriculture. A civilized soil culture works to self-perpetuate, costing the farmer next to nothing. The raw igneous rocks needed by the soil are plentiful and inexpensive. Raw, crushed rock sifted to a 250 mesh costs less than $100 per ton. By contrast, factories producing chemical fertilizers burn excessive energy in production, contribute to acidification of the soil, and are not as readily available to crops as natural mineral supplements. Chemical fertilizers are also a continual, significant cost drain to farmers who choose to ignore the intelligence of their soil.

At Green String Farm, we add finely crushed volcanic rock to the soils to enhance the mineral content, and we add ground-up oyster shells to provide the appropriate calcium levels. These additions help negate the need for pesticides and herbicides. The abundance, health, and consistency of our crops prove to us that land stewardship requires this second crucial food group: rock. The health of our interns and farmhands also sings rock's praises.

3. DANCING DIRT MICROORGANISMS

The elegant ongoing symbiosis of soil life includes the dance of minerals with microorganisms. This dance has many beautiful steps. Acid waste from plants and microorganisms dissolves alkaline minerals, creating new elemental complexes, which in turn power additional growth for plants and soil organisms.

Soil minerals are then made available to complex life forms in an exchange with the compounds found in the atmosphere. This exchange is the work of soil. Rivers flood the lowlands, carrying the biology from the hillside forests to the valley's waiting dirt. Birds and other animals move biology, introducing microorganisms to the fields, inoculating and balancing the soils. Humanity has the mechanical capacity to generate biological teas and apply them to the ground, stimulating the digestion and diversity of soil life.

Thus, life forms in constructive adhesion. It harmonizes, to put it in dance terms. Soil life preceded warm-bodied life, and dirt is the primordial foundation. Listen. It is the underlying music. Fully developed dirt organisms are very specific to their location and moment. Thousands of species, from viruses to arthropods, digest and renew Earth's elements, feeding the plants that gather the sun's energy, creating the opportunity for, and perpetuating the existence of, warm-bodied life. The more life in the soil, the more life it has the capacity to support. That is the music we call life.

4. EATING THE AIR

Just like people, microorganisms have to eat. Their food is fixed air, primarily carbon bonded with hydrogen. We may not derive a nutritious meal from gulping down hefty servings of air, but microorganisms do. They can then get on with the work of mineral digestion.

Plants draw essential elements from the air for growth, as well. A row of broccoli, for instance, begins life as seeds absorbing nutrients from the soil. As each seedling breaks the surface, the plant begins absorbing carbon, hydrogen, and oxygen from our atmosphere. These air and soil elements combine to promote the broccoli's structure, form, and function. Some of the air compounds build plant bodies, while others support soil biology through the root exudates. Upon death, the plant rows contribute these elements back to the soil, feeding its foundation of soil microbes.

It is an elegant symbiosis, which we take for granted. Air feeds soil biology, keeping aerobic and anaerobic microorganisms in balance. Where a highly developed natural soil system exists—supplied with complete mineral availability and copious resources of soil carbon from both living and decaying plants—the soil biology develops to a balanced culture, including free-living nitrogen-fixing populations. Let me emphasize the beauty here: these well-fed underground civilizations draw nitrogen directly from the atmosphere, vastly reducing the need to apply nitrogen to food-producing lands. Perhaps only a farmer sees the value in this. Or perhaps you've heard the news. Commercial agriculture dumps so much acidic nitrogen on the land that most of it is lost, flowing into the water table, poisoning our aquifers, while runoff contaminates coastal marine life. The dead zone in the Gulf of Mexico? Nitrogen runoff from commercial farms. We see the sanity of healthy soils manifest in our fields every year. Our naturally well-developed soils require between two and ten pounds of applied nitrogen to achieve the same yields as conventional crops using between fifty and two hundred pounds.

We must have faith in the intelligence of well-fed fields. We can all step back and let our soils become the productive, air-eating, rock-crunching, microorganism-dancing civilizations that will sustain us for countless generations to come.

CONCLUSION

It makes the oldest sense in the world: foster the four primary food groups of life on planet Earth. Aid civilized digestion, help dissolve the minerals of rock, respect microorganisms, allow the absorption of carbon from the air, and energize the entire, bounteous metabolic process with sunlight.

As farmers, we tend plants as well as dirt. We know how expressive plants are — every physical manifestation indicates conditions of the soil life colony, its scale and civility. May all of us learn to value those expressions and honor what plants are hungry for. We'll have a thriving, well-loved biosphere. We will have found environmental completeness. We will listen as we do a lively two-step with our dirt.

5

NATIVE SOIL

loved and protected?

IT IS THE SORT OF LAND THAT HAS TO
BE ENVISIONED IN TERMS OF WHAT IT ONCE
WAS AND COULD BE AGAIN. THE SORT OF LAND
THAT DESPERATELY NEEDS TO BE LOVED
AND PROTECTED, AND RARELY IS.

DONALD G. SCHUELER
A HANDMADE WILDERNESS

HOSTILE TAKEOVERS

An Ode to Guts and Gardens

✻

LAURA PRITCHETT

he location is beautiful and the soil sucks. I live below the first buckle of foothills of Colorado's Rocky Mountains, waves of mountains to the west and grasslands in all other directions. The only problem? The dirt is as bad as the view is good. Instead of the deep brown rich soil that other locations on the planet naturally enjoy, this soil is composed of sand and clay, but even worse, underneath the thin top layer of soil that supports all the glorious grama and bluestem and sage is, well, rock. Rock and more rock.

It's good that the view is so inspirational, because it has given my family patience during the past ten years of trying to foster soil that was gardenable. Like many people, we've been busy and harried, but we also had a vague but purposeful plan: building soil that would foster vegetables, vegetables that would foster good health.

More or less, this all worked in a haphazard way. Despite the weeds, the drip-irrigation system that often needed revamping, and the persistent plentiful deer, we did, in fact, usually end up with a nice crop of veggies. Pumpkins for Halloween, squash enough to last all winter, cucumbers enough for salads, tomatoes to eat and then sun dry, and, of course, enough zucchinis to give unwilling neighbors.

But then catastrophe struck. It struck in the way that catastrophe often does: with unrelated chaotic events. My catastrophe included a septic tank, a bulldozer, an infected tooth, and antibiotics. From the chaos of all that, I learned to sing songs of praises to microorganisms, both in my garden and in my gut.

It started with an ice storm, which froze the ground, which caused the septic tank to burble out foul-smelling contents in the middle of winter. The appropriate people were called; a new septic tank was in order and would cost nearly $10,000, and, more painfully, would need to be moved because of new regulations. The only place to put it, they said, was where the garden bed now sat.

After years of amending the soil in this huge area, we had finally achieved something that looked more brown and rich and fertile. Having all that dark soil dug up made me consider extreme options, such as never flushing the toilet again. But then, *voilà*! I had the excellent idea to ask the gentlemen to *save* this stuff; they could scrape and bulldoze it to the side or something, right?

But no. Alas.

The crew came while I was gone on a short errand, and before I returned, our hard-earned soil was buried in a huge pit many feet below the surface of the ground. All that good dark brown soil, so far far away.

I stood on top of the rocky "parent material," as it is called, and stared down at it, as if my vision might penetrate and bring up the dark good soil below. The stuff below me had no organic material, no nothing. The ten chickens, who followed me out there, looked up at me. Then they looked down at the crappy soil and looked up at me again. Not even *they* liked this soil. No bugs in it—nothing really to do except take a dirt bath. The golden retriever sniffed it and looked up at me, equally unimpressed.

People who live in such areas will understand my sorrow. All those pickup loads of horse manure, chicken manure, cow manure, any kind of soil we could get our hands on. Kitchen compost, chickens scratching. Even moving worms from the street into our garden with the hopes they would tunnel and loosen the compacted soil (and also just to get them off the street). And now, doing all that *again*?

It made me want to weep.

But I didn't have time, because I also had an aching face. Soon after the septic tank repair, and yes, nearly $10,000 later, the correct two teeth had been identified, root canaled, infected, my gums cut and stitched, infected again, a perforated sinus cavity, gums cut and stitched again. Which required round after round of antibiotics. Which caused my stomach to hurt, because my good microorganisms had been killed too.

My guts were as barren as the dirt outside.

Welcome to the land of hostility, I thought to myself.

When things look bad, I figure, a person can curl up in a ball of sorrow, or they can stand defiantly and shake their fist at the skies. I opted for the second and became obsessed with fixing both. Indeed, these two problems—the barren garden dirt, the barren guts—became related, not only in the sequence of time, but via the very core of the way they function. Indeed, soil fertility and human health are directly related; the word "human," after all, is related to "humus," meaning "earth or soil."

As I discovered, we humans are just beginning to understand exactly what that means. Both the human gut and the layer of dirt are in some ways still unexplored territory. It's complicated in there! There are one hundred trillion bacterial cells in a human body, ten times the number of human cells. About five pounds of your body weight is bacteria. There are one hundred million to one billion bacterial cells in one teaspoon of healthy soil. Which reminds me of Leonardo da Vinci: "We know more about the movement of celestial bodies than about the soil underfoot."

Acclaimed science writer Michael Pollan's essay "Some of My Best Friends Are Germs," for example, discusses the role of microorganisms in our guts and the new science that suggests that autoimmune diseases are related to a disruption in the relationship between our bodies and their "old friends." Like garden dirt, our guts need microorganisms. To summarize a long complicated ode to these little bugs, I'd just say that due to our farming and eating and kitchen practices, we aren't getting that good broad spectrum of bacteria anymore. Also, these little bugs need upkeep. Part of *that* has to do with soil.

So first, I turned my attention to the soil. As someone who is always a little too busy and a little haphazard, I have never quite known what exactly I am supposed to be doing to my garden dirt. I've never been one of those organized people who had the pH balance measured or whatever hyper and vigilant people do. But I know instinctually that garden dirt is supposed to be brown and rich and full of nutrients and soil microbes and organic matter. I know when my tomatoes don't come up that something is wrong (okay, so sometimes it's the deer. But more often, it's the soil).

Although I may not know the specifics, I have pretty strong opinions about the general gist. Stewarding good soil is, I think, one of our jobs as humans. Fostering humus is fostering humanity. The more humus, the more fertile the soil is. Indeed, humus is *the* most critical aspect of soil composition. It's that spongy, amorphous substance and is basically organic matter. If I closed my eyes and pictured a forest floor in Oregon or something—all those layers of fallen leaves and plant and animal debris falling and decaying—that's humus. Humus is considered by some scientists to be the greatest water management and conservation tool available to us. Humus protects plant roots from drought, salt, temperature, and disease, and it also radically stimulates microbes. And that is not what I had.

So we started again. Food scraps out to the compost bin, compost bin turned and watered, compost spread. Sometimes, though, I just cheated—and I have discovered this is not such a slobby practice after all—and I took nonmeat food scraps straight out to the garden and dumped them. Also, we brought in good ol' manure from the family ranch. The chickens scratched and scatted. Slowly, over the course of winter, and into spring, things started to decay. We rototilled, threw more food waste on, rototilled again. By spring, that parent material looked at least semi-gardenable.

Meanwhile, I was trying to help out the little bugs repopulating my guts. The microorganisms in our bodies can use a little help too. They do more than digest food; the little creatures in our digestive tract play a role in our immune systems and overall health. This too is unexplored territory, and science is just now beginning to understand the importance of the creatures that live inside us.

One fascinating fact I discovered from Pollan? Our gut bacteria play a role in the manufacturing of substances like neurotransmitters, enzymes, and vitamins (including important amino acids and short-chain fatty acids). But more than that, they help regulate our stress levels and even temperament. As Pollan points out, when gut microbes from easygoing, adventurous mice were transplanted into the guts of anxious and timid mice, the mice became more adventurous. Also, here's another cool fact: it's the gut microbes in locusts that cause locusts to swarm! The microbes release chemical signals that cause the locusts to physically change in shape and develop a radical new behavior. The expression "thinking with your gut" may contain a larger kernel of truth than we thought.

As per usual, I wish I were one of those vigilant people who was organized and kept track of everything she put into her body. But no such luck. I figured if I haphazardly put good stuff into it, all would be well, more or less.

But as we all know, personal crises can induce change. And fast. As I moved from Clindamycin to a Z-Pak to others, pus pockets formed on my gums, and the seriousness of my predicament started to grow. My stomach *really* hurt as the yeast colonized and conquered. As my microbes died. I didn't really know *what* was going on in that dark cavity inside, but I did know how much stomach woes really are painful and life limiting.

The scariest part of this war was that I learned about the long-term consequences of antibiotic use. One course of antibiotics may bring shifts in the relative population of the various species of bacteria in our guts, but usually the guts can recover. Several rounds of antibiotics, though, can change your gut *forever*.

In extreme cases, some individuals need a "fecal transplant," which involves, yes, getting someone else's poop into a sick person's gut. *No thanks!* That poop transplant motivated me to nurture the health of my microbiome. So I got serious. And I got simple, obeying the most basic truth about healing the gut: to heal gut flora, you have to repopulate your system with good bacteria and return the gut to pre-antibiotic levels of beneficial microbes. My mission to repopulate my guts began with figuring out how to do just that.

Basically, I eliminated some foods and ate a lot of others. The first things to get rid of, I discovered, are sugar and processed foods. And the first things to eat? Yogurt and other foods with prebiotics and probiotics (the pre- being the more important of the two). I ate garlic and lemon, because I like them. I ate beets and sauerkraut, even though I don't.

And, yes, I went back out to the garden. I picked out the rocks, turned the soil. As soon as it was warm enough, I planted unique things for me (Artichokes! Not so easy to grow in Colorado!) and all the usuals (Ah, zucchini). I planted marigolds to keep out the pests. I transplanted raspberries to new patches. Started a new grapevine. Put in new peach, apricot, cherry, and apple trees. Tried again for strawberries (but those deer just won't leave them alone).

Luckily for me, gardening is a good idea for guts. Fresh veggies and dirt have the little microbes we need. Indeed, some health care professionals argue for relaxing the sanitary regime in homes, encouraging folks to wash their veggies less and spend more time in dirt and with animals, thereby deliberately increasing their exposure to the grand patina.

In the end, the patina solved both problems. And pretty quickly, too. Within a few months, the garden soil was strong. Of course, we'll need to keep on amending the soil, but it wasn't the huge chore I'd feared—a few months rather than a decade. It was not so hard to parent the parent material after all. And this good soil had a cascading effect. My chickens wandered around the yard, for example, looking sun-drunk, more pleased with the bugs in the dirt, and their eggs tasted even better.

My health, too, rapidly improved. The tooth problems eventually got solved and the stomachache beneath my breastbone quit hurting. And as soon as one feels good, well, that leads to a cascade of good things too. More smiles, more energy, more plants in the garden.

But while the body and garden are back to normal, I'll never quite be the same. For a mere $20,000, I sometimes joke, I've had a crash course in the deep appreciation of the power of microbes to do their dance of health. I lost my obliviousness to the ways in which I should always nurture both. (Even those who are a bit rock-headed can become as soft as humus!) I now feel it in my bones: we are from the soil and we'll go back to the soil. And while here, we can work to steward both.

Now that it's summer, I've been standing in my garden and watching the sun set over the blue waves of foothills in my little valley. The garden has offered up plenty. I've been drying, freezing, and eating a lot of fresh and beautiful and colorful veggies. Peaches, apples, cucumbers. My fridge and freezer are full, my belly is full, my life is full. The deer sometimes walk by, and pause, and watch me watching them. A bear has been spotted wandering by. The golden retriever pounces around the yard, leaping at the occasional falling yellow cottonwood leaf, and the chickens wander happily. All of it mixed together—the view, the garden, the body, the life—strikes me as the grand patina, the great palette of color and mystery. A territory no longer hostile. A home.

FIGHT THE POWER

❋

E B A N   G O O D S T E I N

*L*eave the bustling border town of Santa Elena: Venezuelan/
Brazilian *chriolla* of *tiendas*, bodegas, car repair shops, super-
mercados stuffed with global brands. On the streets, beautiful young
people, all curves and planes, Möbius-like, each a self-contained
universe. The middle-aged, relaxing into their round shapes. Elders,
smaller, moving slowly, from a different time. The whitewashed concrete and adobe walls, along the *calles* and out onto the highway, all
boldly stenciled over with the identical post-Chavez election concept:
M♡duro, creating a visual unity to the built landscape. Pass the towering, martyred image of Elena herself, watching over the road, and
drive north into the Gran Sabana.

This feels first like high desert country: Colorado, Montana, eastern Oregon. Vast rolling plains of grass, big skies, distant mesas,
narrow groves of trees running up the draws and valleys, telling of
reliable water. Only with scattered green palm trees. And then a roadside attraction. An adobe hotel shining in the sun, thatched roof, set
on a bank at a place where a wide, slow-moving river suddenly col-
lapses, acre-feet of water dropping two hundred feet in four seconds,
a roil of resistance and inevitability.

After an hour, stop at San Francisco. The sky has darkened quickly,
and, Coke in hand, sit under the tin veranda and smell, then hear,
then see, then feel, the deluge as it rolls in. In minutes, the packed
clay parking lot is swirling in an inch of water, rivers everywhere. The
symphony lasts ten minutes, a magnificent entertainment, and twenty
minutes after that, save for a few puddles, the land is dry once again.

From San Francisco, drive twenty kilometers east up through the
sabana, along a wide clay road, the last stretch steep and rocky. No
palm trees anymore, just grasslands. Pass a dozen children in uniform, assembling outside the freshly painted school, and enter the

village of Paraitepui. The final destination is three days' walk from here: Roraima, one of the oldest places on the planet of Earth.

As Gran Sabana rises steadily from the river basins, it is capped in places by *tepui*, mesas kilometers in length, surrounded on all sides by sheer, impenetrable, thousand-foot cliffs of layered sedimentary rock. Stream waters cascading from the plateaus fall farther here then anywhere else in the world. *Tepui* are iconic, familiar places in the cultural mind of the West. They were first Arthur Conan Doyle's *Lost World*, in turn the inspiration for Stephen Spielberg's *Jurassic Park*, recently the destination of an old man and a Boy Scout in Disney's *Up*.

Among the *tepui*, Roraima is uniquely accessible. A natural ramp of broken rock on the south face provides a route to the summit. Westerners first ascended and described Roraima in 1884. Thirty years passed before Conan Doyle's fictional explorers discovered fictional dinosaurs still alive at the top. But this lost world derives from an age much, much older than those of any saurian. The *tepui* are the remnants of a once-connected landscape stretching thousands of kilometers across northeastern South America and southwest Africa.

About two billion years ago, when most of Africa and South America were joined together in the single land mass of Gondwana, a mighty river flowed across the land, emptying into an ocean basin, like the Mississippi into the Gulf of Mexico. The world was quiet then: multicelled life was not yet, and would not be for more than a billion years. Silent save for the wind and rushing water, over eons, fine-grained sand and muds poured into the basin and onto the underlying igneous bedrock, forming layer upon layer of sediment, thousands of feet thick. As geologic time passed, these deposits were in turn buried under thousands more feet of additional sediment. Deep underground, the intense heat and pressure led the layers of silica and clay to crystallize, forming a massive layered slab of the hardest sandstone.

In this way, four hundred million years passed. And then, the earth's first mountains were built. The deeply buried *tepui* sandstone was thrust vertically upward as a single block, retaining the flat, horizontal structure of its original rock bedding. The Roraima formation became the heart of a new mountain range thousands of meters high.

And since that age, so long ago, nothing geologically dramatic happened to the *tepui*.

Only this. Another billion and a half years of rain and wind have leveled the landscape, carrying most of the mighty mountain range, rock by rock, grain by grain, down through today's Orinoco river basin, north to the sea. In most places, time has succeeded in peeling back the thousand-foot layers of quartz sandstone and grinding down the older igneous rocks below, leaving only a thin layer of soil over the rolling plains of the Gran Sabana.

But at Roraima and Canaima, and a few dozen other sites across Venezuela, Guyana, and Brazil, the *tepui* have resisted, and still they rise.

Traditionally, the indigenous people, the Pemon Indians, stayed away from Roraima. Their story is that the mountain was the stump of an ancient tree, once holding all the fruits and vegetables in the world. But it had been felled by a trickster, leading to a catastrophic flood. Today many of the residents of Peraitepui travel regularly to the top of the mountain as guides, cooks, and porters, carrying food, tents, and other supplies for the six-day trek. They ask, however, and expect that visitors will treat the mountain with respect.

From the scattered homes in the village, Roraima and her sister *tepui*, Kukenan, rise fifteen miles to the east and north. Kukenan has no natural route of ascent and is off-limits to visitors; the second highest waterfall in the world falls from the south face. This is the beginning of the rainy season, June rain every day in the afternoon, heavy dew at night, mist. Clouds hang permanently between the two *tepui*, and the tops of the mesas are also shrouded, rarely seen. Flow elevations in the rivers vary hourly with the rainfall. No mammals evident, only lizards, spiders, birds, ants. No mammals, no giardia. Drink the cool, fresh water in these streams, the rain falling from Roraima and Kukenan.

On the two-day walk to the base of Roraima, the paradox of the grassland presses. Unlike the desert rangelands of the American West, this is a place of abundant rainfall: the tropics. Why is the grass so sparse? Why do the trees cling to the streamsides in the valleys? Why, in short, isn't this place a rain forest?

Because the soil is thin and poor.

Across the plains, the trails through the grass are almost paved by smooth red clay, a hard rain barely penetrates the surface. This is laterite soil. The force of the tropical rains leaches the organic matter, leaving little behind but clay. The underlying, ancient bedrock, the source of the soil, is acidic, an impoverished substrate. And unlike the American prairie, or the African savannah, no vast herds of buffalo or antelope roamed these grasslands, transforming and redistributing nutrients and breaking the ground under their hooves. The Gran Sabana, simply, is not an easy place for dirt.

A hundred miles south of here begin the vast rain forests of the Amazon, where a different balance has been struck between tropical rain and earth. As in the river valleys of the *sabana*, and from comparable soils, forests slowly expanded their reach. The initial forest cover created some protection from the drenching rains, as well as root structures to hold and enrich the nutrient-poor soils, enabling in turn the slow accumulation of organic matter. The trees themselves create a microclimate that tempers the dry season and recycles moisture. All this supporting further forest growth in a virtuous cycle, but at a tenuous balance.

When tropical rain forests are clear-cut for agriculture—their laterite soils exposed again to the full leaching power of the rain—they can desertify, flipping forest to savannah. Over the coming decades, as global warming heats up and dries out the Amazon, the forest will likely burn, clearing much of the land. And then, Gran Sabana —already one dominant biome of the laterite—may become even grander, imperial, perhaps triumphant, spreading deep into the Amazon basin.

Rising above the *sabana*, today, a strip of rain forest rings the base of the Roraima cliffs, supported by the perpetual mist from the mountain. Climb up through this forest, reach the cliff wall, touch the rock face, look up at a summit lost in clouds. Work up the ramp, pass under the constant rain of tears, Las Lagrimas, cascading a thousand feet from the summit. Glancing back, in the sunlight, a rainbow. Pause by a flat rock, the white and brown rippled surface etched by the waves of a two-billion-year-old sea. Topping the last boulder, reaching the plateau, it is as if you have rounded an unexpected corner and stepped onto the moon.

With moon plants. There is a well-tended garden with tall, spiky yellow flowers, and low, red-barked trees arranged carefully among the rock crevices. Around the next bend, confront a massive sculpture. A purely symmetrical, oblong, black stone, ten meters long, three meters high, balanced on a thin pedestal, rising from the same black rock. It has a name: La Tortuga.

On Roraima, each stump of rock is worked by uninterrupted time, endless years of hard rain, mist, steady wind, and the drag of gravity. A slight acidity of the water seeks the more basic layers, dissolving bonding agents in the rock. Imperceptible forces make a sculpture garden of the shallow cliffs and boulders rising from the flat plain of the mesa. Weird figures emerge, like those featured on roadside billboards advertising tourist caves and caverns: "See the Mother and Child!" Here, the Camel, the Dragonback, the Valley of the Penises. Roraima is a cave floor, kilometers in length and breadth, only with no overburden, fully exposed to the stars.

The surface of the plateau is hard, bare rock, with a darkness the stain of a lichen. The rocks are rounded, latticed by tight crevices, everywhere solid and smooth. Nothing rotten, every step secure, like walking through mist on the impenetrable back of a vast, ancient turtle. Down the low rises, vertical veins of quartzite sometimes break the shell, with piles of crystals spilling down to the base. Here and there are accumulations of yellow mud and standing waters in the shallows, and in holes and crevices, a thin soil, supporting the work of the gardener: thirty types of orchids, endemic insectivore plants, dwarf-kin of agave.

Not far from La Tortuga, find the cold baths. Four perfect Jacuzzis of transparent rainwater. Cool, neck-deep when seated, bottom and toes settling delightfully into a pure floor of broken quartzite crystals. Below the polished rim of each tub, in a trick of the light, almost turquoise, the smooth pool walls reveal a flesh of reddish stone under the dark shell.

There is a fierce wind and driving rain, but it never freezes on the mountain. No extremes of weather. For more than a billion years, this mass of rock has resisted, year in and year out, the steady, patient action of nature and gravity, seeking to turn rock into soil, to make a place for the roots that could break this mountain and drag it back to the sea.

Kneel by a small sculpture. The rain, the mist, the wind, have preferentially eaten away at a lower rock layer with a softness barely perceptible to chemistry itself, creating a rounded overhang. Under the overhang at the very back of the ledge, out of the wind and rain, some moss has taken root. At the base of the moss, a small red flower grows from a tiny patch of soil.

The oldest dirt in the world.

BORN AGAIN

Loving the Least Worst Land in Mississippi

from *A Handmade Wilderness*

※

DONALD G. SCHUELER

*T*he dream that Willie and I had was to own land in the country. But not just any land. Not, for example, some cute little cabin in a resort development, and most definitely not the sort of "secluded" three-acre lot in the wilds of exurbia that is presently gobbling up even more of rural America than suburban sprawl itself. In its specialized way, our dream was more ambitious than that. Before we got together, Willie had never seen anything greener than New Orleans's City Park in his whole life, but the idea of "the country" entranced him. He wanted to grow things, and he loved all shapes and sizes of animals. It thrilled him that he might have as neighbors creatures he had only seen in the Audubon Zoo or on TV nature shows. Also, he had a taste for living adventurously. As for me, I was an amateur environmental activist who loved the natural world and wanted to protect it and live close to it, at least on a part-time basis. Together, our enthusiasms added up to a great yearning to have our own private nature reserve, a place that would be a haven, not just for us, but for the flora and fauna that would share it with us.

What we needed, obviously, was a pretty sizable chunk of countryside; enough space so we could be in touch with the natural world without moving it out when we moved in. But when we started our search, in 1968, we didn't know how much space was enough, or where to look for it. Or, most crucial of all, how much we could afford to buy. Willie was a floor sander and I an assistant professor at a state university. Which is to say that we could be classified, at best, among the country's Taxable Poor. Moreover, like all good Americans, we were in debt. We had just finished fixing up a neglected house in one of New Orleans's many wonderful old neighborhoods, and the bills were still coming in.

It took us a while but we eventually concluded that we were not going to be able to afford a large slice of some scenic natural wonderland. In the Deep South, as in most of the more heavily populated regions of the nation, pristine wild land is almost nonexistent, and what little is left ought to be, and sometimes is, protected from any kind of human settlement. But, guided by real estate agents, we did find some large tracts of private land for sale that would have suited us just fine. Places that were within a couple of hours' drive of New Orleans, furnished with magnolias and mossy oaks, and usually situated along the banks of some lovely stream or bayou. But you pay extra for rustic scenery. At the going prices, the amount of land we could afford would have been barely enough to accommodate us, a pair of mockingbirds, a few squirrels, and maybe half a raccoon.

It took us a couple of months of that sort of window shopping before we decided to face reality. We could buy some tiny bit of real estate that already looked the way we wanted it to look, or we could (maybe) buy a decent-sized piece of land that didn't; but we couldn't have both. From that realization evolved another: that land, like the half-wrecked old house we had restored in the city, could be rehabilitated. There was, however, one big difference: a piece of land, unlike a house, would do most of the restoration work itself if we just gave it enough time.

We adjusted our expectations accordingly. We would settle not just for any second- or third-best land, but for as much of what Willie called "the least worst land" as we could afford to buy at the cheapest price. No doubt a certain measure of sour grapes went into that decision, but we were excited by it, too. We had posed ourselves a challenge and then taken ourselves up on it. Whatever else, we were in for an adventure!

There are millions of acres of least worst land available in this country, much of it within reasonable commuting distance of large cities. It is land that wears the heavy scars of human abuse, land that is not near coastal beaches or pretty inland lakes or fashionable ski slopes — land that the developers wouldn't touch with a ten-foot surveyor's pole. It is the stony gray hills of the Northeast, covered with thin third-growth woods; the overgrown pastures and fields of abandoned farms in the Carolinas; the ravaged sites of former strip mines in the Alleghenies; the unirrigated prairie corners of the Midwestern corn belt; the near desert of expired sheep ranches in west Texas and New Mexico; the dry scrublands east of the Cascades. It is the sort of land that has to be envisioned in terms of what it once was and could be again. The sort of land that desperately needs to be loved and protected, and rarely is.

Here in the Deep South, land like that is almost always composed of pinelands or hardwood bottomlands from which virtually all the marketable timber has been stripped. There are thousands of acres of that kind of cutover woodland available for sale at any given time, but Willie and I had a hard time finding a piece of it that was right for us. The real estate agents on the Mississippi Gulf Coast never did catch on to what we wanted. They were bemused when we said that we actually preferred land that had some swampy ground on it, that we didn't care if the timber had been recently cut, that we definitely didn't want frontage on a paved road. And no, we weren't interested in a property with a dwelling on it, unless it was an old country homeplace we could fix up ourselves. Finally, they and we gave up on each other with mutual sighs of relief, and we went land hunting on our own.

What we did first was buy U.S. Geological Survey maps of areas that, judging from ordinary road maps, seemed to be not too heavily populated and at the same time not too far from New Orleans. They were incredibly detailed, indicating not just land elevations and main roads, but the location of every dwelling, pond, streamlet, and dirt track to be found in a given quad at the time the map was drawn. With these for guides, Willie and I, in the company of Schaeffer, our fawn Great Dane, and Sammie, our thoroughbred mutt, roved the back roads of southwest Mississippi in our Volksbus for half a dozen winter weekends in a row. We checked out faded For Sale signs nailed to fence posts at the edge of weedy pastures and burned-over woodlots. We accosted local people in the front yards of their trailer homes, asking them if they knew of anyone roundabout who was trying to unload some property. Sometimes all we got was a suspicious look that passed from my white face to Willie's black one, but sometimes we were rewarded with a name and directions to someone's house.

During those weekends, we tramped a lot of parcels of land. We were like Goldilocks: even the most likely properties didn't quite fit; they were too large or too small or too expensive or too close to the neighbors or too ecologically bland. Yet all along we felt that we were getting warmer, that the just-right property was out there, waiting for us, and sooner or later we would connect with it.

Meantime, we were learning just what least worst land meant in practical terms, and what it cost per acre. Even more important, we were getting some idea of how much land would be the minimum needed to create the sort of pocket-size nature reserve we had in mind. You don't have to be Aldo Leopold to figure out that forty acres of mesquite in south Texas isn't going to give you as much biodiversity as, say, a small, well-watered dale in southern Pennsylvania. On the other hand, you can probably buy half a dozen forties in south Texas (minus mineral rights, of course), for the price of one in the Northeast. Indeed, in that open country you would want to anyway, to keep your nearest neighbor from shooting at your jackrabbits and coyotes from his front porch.

Willie and I eventually decided that one south Mississippi forty might, just barely, serve our purpose. As it turned out, however, we ended up buying twice that much. What happened was that one chilly, raining day in December, we were out questing once again, and we ended up on a particularly wet, muddy dirt road that snaked this way and that in its not very successful effort to avoid the many narrow creek bottoms that tried to intercept it. Now Volksbuses are not noted for their traction in slippery situations such as this, so I suppose I shouldn't have been as surprised as I was when ours embarked on a long slide while I was trying to get around an especially sharp curve.

"Uh-oh," said Willie, pressing his hands against the dashboard. Sammie, easily rattled, hopped into his lap. Even Schaeffer, usually the most imperturbable of dogs, looked concerned.

For a few seconds everything was out of control, with the Volksbus skating perilously close to the edge of a deep drainage ditch. Then the road, though not the Volksbus, straightened out, bringing into view two men standing right in the middle of it. To judge from the expressions on their faces, they had never seen a Volksbus driven sideways before. They both carried shotguns, and their first impulse, as we swooped down on them, was to point the guns in our direction rather than jump aside. Luckily, the bus came to a halt—still facing sideways—a few feet from where they stood ankle deep in mud.

"Hi," I said, giving them my very best effort at a disarming smile.

They didn't reply, but they did lower their guns.

All things considered, they were not the sort one would want one's sister to date. Low brows, sullen eyes, paleolithically slouching shoulders. For several seconds they just stood there, slowly sizing us up. No doubt we did seem an unlikely quartet to be roaming the backwoods of Mississippi: Willie, trim, handsome, black; Schaeffer, large even by Great Dane standards, pressing his dark mask against a back window; Sammie, barking ferociously from the safety of Willie's lap; myself, flummoxed and no doubt citified-looking, trying hard to think of something folksy and friendly to say.

On Schaeffer's account a bit of light finally flickered in the men's eyes. Evidently they had never seen a Great Dane before.

One of them asked, "What's he good for?"

"Well," I mumbled, "he's, uh, sort of a pet."

The light in the two pairs of eyes went out. I might as well have told them he was an objet d'art. "Of course," I added defensively, "he's a very good watchdog." Which was not exactly true. Then, to change the subject, I asked them what they were hunting.

"Deers."

"That's great," I said. I meant it, too. It wasn't that I had any enthusiasm for deer hunting, but I was glad to learn that there was a population of deer hereabouts that was numerous enough to hunt. I had hoped there might be; but there were also a good many humans living along the back roads we were exploring and, just judging from first impressions, I suspected that even the adaptable whitetail might find it hard to coexist with them.

"Any luck?" I asked.

"Naw," they said.

As though to confirm his observation, the air suddenly turned loud with a fanfare of complaining yaps and yelps. The next instant the van was surrounded by a swirling pack of black and tan deer hounds. They ignored Sammie's yammering, but when aristocratic Schaeffer, confronted with this canine rabble, uttered a contemptuous "Woof," they did their excited best to climb through the windows.

The spectacle was too much for Willie. "They don't like him being so high and mighty," he laughed. And when Schaeffer woofed again, even the two men had to grin. "He do look like a deer," one of them said.

"You're right. Ha, ha," I answered, beginning, on that benign note, to carefully back the Volksbus around in the midst of houndish chaos. I figured we had had enough country motoring for one day. But when at last we were about-faced, I asked the hunters the question I had asked so many times during these last weeks: "You wouldn't know if anyone has some land for sale around here, would you?"

Pause. Then one of the men answered, "Well, you might could try Old Man Stanton. Lives about five miles south on the hardtop. He's got a bunch a pieces all out in here."

We did try Old Man Stanton, who turned out to be a spry, round little gentleman in his seventies, with a habit of appending wicked little chuckles to everything he said. And, yes, he had a few pieces of land in the neighborhood that he had acquired during the Depression as a payment on defaulted loans. Right off he seemed to understand what we were looking for in the way of bargain-basement property, perhaps because that was the only kind he owned. He told us to come back the following Saturday. When we did, he donned old-fashioned army leggings ("Snakes, y'know, heh, heh") and energetically led the way on a day-long progress during which we toured a forty here, then an eighty over there, then another forty "just down the road a little ways, heh, heh."

By late that drizzly afternoon, after we had slogged across what must have been miles of beige meadows, plundered pine woods, and flooded gullies, even Mr. Stanton admitted to being "a mite give out."

The supply of his scattered holdings was giving out too. Wearily, we climbed out of the car to have a look at the last property, an eighty, that he had to show us.

It was the Place.

God knows, our Promised Land didn't look all that promising on that dreary December day. What it did look was *big*. To a couple of city rubes, a tract one-half of a mile long by a quarter of a mile wide seemed a vast expanse of real estate. As the buzzard flies, it lay about twenty-five miles back from the Gulf Coast in a relatively narrow belt of gently rolling land known locally as the sandhills. During the Pleistocene epoch, between 1 and 1.5 million years ago, ancient rivers shaped its modest ridges and hills out of the alluvial clay and sand they had carried with them in their southward progress. Whether the wilderness that covered those sandhills was overwhelmingly dominated by towering pines or Sherwood Forests of massive oaks, it must have been something really wondrous to see.

In terms of human exploitation, these forests were, and still are, the only important natural resource that the sandhills can boast. The thin, sandy loam topsoils, highly acidic and nutrient poor, are ill-suited for anything but subsistence agriculture; and no mineral wealth has been found beneath the underlying hard clay pan. This poverty of resources explains the poverty of the sandhills' human history. The Choctaws used this area only sporadically, preferring the richer hunting grounds of the alluvial river bottomlands that intersect the sandhills to the east and west. Since the 1840s, when the tribes of the Choctaw Nation were deported to Oklahoma, and under the none-too-gentle management of folks like Mr. Stanton, much of the Place and the land surrounding it had been cut over. Whenever a stand of pines got tall enough to be converted into broomsticks, pulp, or creosote fence posts, down it went. Meanwhile, the local people — unimpressed by the property rights of absentee landowners — set the hills afire every spring to eliminate dry winter grass and pine straw, thereby inducing early grazing for their free-ranging herds of half-starved cattle.

This, then, was the much-exploited landscape that Willie and I were ready to embrace as our own private Shangri-la. Right off, we loved it. Partly because we were already seeing it as it could be. And partly because, in the way of stray dogs, it looked as if it could use some tender loving care. I confess, in those early years, I sometimes stretched out on my stomach in the leafy mold, courting chigger bites, but also in some primitive way inviting the familiar maternal earth to lend me what strength it could.

We were so lucky. We couldn't know it at the time, but those first eighty acres contained a microcosm of virtually all the ecosystems that the Mississippi sandhills had to offer. To be sure, it was all battered, chewed up, scorched, and generally much put upon; but nearly everything was still at least embryonically in place. Given half a chance, it would show a Southern Baptist a thing or two about being born again.

In 1992, Willie and Don's handmade wilderness became the Willie Farrell Brown Nature Preserve, which is overseen by the Nature Conservancy.

STEWARDS OF THE LAND

from *Nature as Measure*

❋

WES JACKSON

*O*ne recent June Sunday two friends and I were driving home along a blacktop road through south-central Kansas, Mennonite country. The previous night, and continuing into the morning, much of the state had experienced hard rains, in some places five inches and more. Such storms are not infrequent in Kansas. From earliest childhood, native Kansans, indeed all Great Plains people, are keenly aware of the hair trigger which stands between drought and instant drenchings, which are often accompanied by spectacular lightning displays and high winds.

As we drove through this relatively flat and prosperous Mennonite country, with its tidy fence lines and well-kept houses and farm buildings, we saw roadside ditches and newly opened furrows. They probably contained milo sorghum seeds, but were so full we could not tell whether the crop had germinated or not. The ditches and furrows were not full of water, but of rich black mud—that blackness characteristic of fertile prairie soils. This particular landscape has little topographic relief, but what little there is had been accentuated in the past few hours by diagonal washes, five feet wide and more, leading down to the ditches which were now level with the adjacent fields. The little streams of the area were running full and muddy.

A hundred years ago the German-speaking, Russian-born ancestors of these Mennonites had introduced hard winter wheat to the United States, and with it the easily copied cultural practices that eventually gave the Great Plains region its well-deserved reputation as a breadbasket. These farmers, like their close religious relatives, the Amish, believe the highest calling of God is to farm and be good stewards of the soil. Within an agricultural context, they are usually regarded as the most ecologically correct farmers of any in America. The strong ethic of land stewardship is, without a doubt, largely responsible.

Less than an hour's drive to the east, my friends and I had spent a memorable, leisurely afternoon surrounded by several thousand acres of tall grass prairie country in our state's lovely Flint Hills. We had met other friends and together had botanized, birded and picnicked under a still cloudy but unthreatening sky. Upland plovers were everywhere joining their sounds with the nighthawks, scissor-tailed flycatchers, and meadowlarks. The storm seemed to have immediately invigorated such attractive plants as Showy Evening Primrose, Pale Echinacea, Plains Larkspur, Butterfly Milkweed, and Lace Grass. It was clear on this rich prairie that the rain was being retained long enough in the spongy mass to give the soil a chance to slowly soak it in and then become a reservoir of water for future needs.

In the Mennonites' field, the water had run off, except where it stood idle in puddles. Soil that had become mud was deathly quiet. Even the most casual observer of nature would not fail to see the contrast. The hills are living, and, so long as they are clothed, eternal; the relatively flat lowlands, put to the plow by scarcely three generations of land stewards, are ephemeral.

As we stopped to photograph one severely eroded field, my mind turned to the owner of that field and tried to imagine what was going on in his head on this wet Sunday. First of all, he will have to replant. It will be substantial, he will think, but necessary and affordable. But in this late afternoon, before chore time and evening church services, is he wondering how many more rains like that his fields can take, and is he asking what, after all, is the meaning of land stewardship, which is central to his faith?

A Mennonite at a rodeo is unlikely. Rodeos are wild places; the boisterous sons of ranchers are not known for their piety. But these cattlemen are stewards of the grasslands—though they probably don't think of themselves in that language—and they need only one ethic. It is simple, straightforward, and easily taught to their children: "No more than one cow-calf unit to about seven to ten acres, and start moving them off when it's dry." The rancher knows it can rain and blow, and his sons can attend the rodeo, chew tobacco, drink beer, miss church, and never mention stewardship, let alone think about its implications. Many ranchers do overgraze and their soil does erode, but even with overgrazing, poor ranching, and no ethic, the land fares generally better when grazed than when put to the plow.

For some soil types, under some climatic conditions, a strong stewardship ethic works — but this is the exception rather than the rule.

In the earliest writings we find that the prophet and scholar alike have lamented the loss of soils and have warned people of the consequences of their wasteful ways. It seems that we have forever talked about land stewardship and the need for a land ethic, and all the while soil destruction continues, in many places at an accelerated pace. Is it possible that we simply lack enough stretch in our ethical potential to evolve a set of values capable of promoting a sustainable agriculture?

WE ARE SOIL

✽

VANDANA SHIVA

We are soil. We are Earth. We are made of the same five elements —earth, water, fire, air, and space—that constitute the Universe. What we do to soil, we do to ourselves.

And it is not an accident that "humus" and "humans" have the same root.

This ecological truth is forgotten in the dominant paradigm, because industrial agriculture is based on eco-apartheid, the false idea that we are separate and independent of the Earth, and also because it defines soil as dead matter. If soil is dead to begin with, human action cannot destroy its life, it can only "improve" the soil with chemical fertilizers. And if we are masters and conquerors of the soil, we determine the fate of the soil; soil cannot determine our fate.

History, however, gives witness to the fact that the fate of societies and civilizations is intimately connected to how we treat the soil—do we relate to soil through the Law of Return or through the Law of Exploitation and Extraction? The Law of Return, of giving back, has ensured that societies create and maintain fertile soil and can be supported by living soil over thousands of years. The Law of Exploitation, of taking without giving back, has led to the collapse of civilizations.

Contemporary societies across the world stand on the verge of collapse as soils are eroded, degraded, poisoned, buried under concrete, and deprived of their life. It can go differently. Twenty years ago I started the Navdanya farm in Doon Valley on a piece of land left barren and sandy by a eucalyptus plantation. Eucalyptus fits into the ecosystems in Australia. But in India it cannot participate in the Law of Return. Its leaves do not degrade, it releases allopathic terpenes that prevent any growth of other plants, and it takes up too much water. There were no soil organisms, and the soil had no water-holding capacity. With love we grew diversity and gave back as much of the organic matter to the soil as possible. Today the soil is thriving with organisms, the earthworm molds cover the farm, we have been able to reduce water use by 70 percent because the soil can hold water, the soil smells with life and gives us life. The other day we had a feast of uncultivated edibles, biodiversity that grows spontaneously because it has found its ecological niche as the soil's fertility has returned.

Industrial agriculture, based on a mechanistic paradigm and intensive use of fossil fuels, has created ignorance and blindness to the living processes that create a living soil. Instead of focusing on the Soil Food Web, it has been obsessed with external inputs of chemical fertilizers — what Sir Albert Howard called the NPK mentality. Biology and life have been replaced with chemistry.

This cannot be our wisest course of action.

Chemically created external inputs and mechanization create the imperative for monocultures, because external inputs can match only one crop, and mechanization requires uniformity. Internal input systems are based on self-organized, mutually supportive diversity. On a small farm, diversity is not an obstruction, it is a celebration of community and coevolution. By exposing the soil to wind, sun, and rain, monocultures increase erosion by wind and water. Organic matter is depleted. Soils with low organic matter are also most easily eroded, since organic matter creates soil aggregates and binds the soil.

We must realize that due to intensive plowing and monocrop plantings, 75 billion metric tons of fertile soils are lost from world agricultural systems each year.

India is losing 6.6 billion metric tons of soil per year.

China is losing 5.5 billion metric tons.

The United States is losing 3 billion metric tons.

Soil is being lost at ten to forty times the rate at which it can be replenished naturally. Soil nutrients, lost to erosion, cost us $20 billion annually. Chemical monocultures also make soils more vulnerable to drought and further contribute to food insecurity. This implies 30 percent less food over the next twenty to fifty years if we rely on industrial agriculture to feed us.

But no technology system can claim to feed the world while it destroys the life in the soil. This is why the Green Revolution or genetic engineering's claim that they will feed the world is false. Intrinsic to these technologies are recipes for killing the life of the soil and accelerating soil erosion and degradation. Degraded and dead soils, soils without organic matter, soils without soil organisms, soils with no water-holding capacity, create famines and a food crisis. They do not create food security.

We have to feed soils on the basis of the Law of Return.

I grew up on my mother's farm in the Himalayan foothills and played with dirt, knew the smell of humus, and loved the smell of the first rain mixing with the life in the soil. While doing research on the Green Revolution in Punjab for my book for the United Nations University, *The Violence of the Green Revolution*, I looked for books about pre–Green Revolution agriculture in India.

I found a copy of *An Agricultural Testament* by Sir Albert Howard.

Albert Howard was sent to India by the British Empire in 1905 to improve Indian agriculture with chemicals. He arrived and found the soils were fertile and that farming was as perennial as the forest. He decided to turn the Indian peasants into his teachers and wrote *An Agricultural Testament* based on learning from nature and farmers.

Howard's work showed that "the foundations of all good cultivation lies not so much in the plant as in the soil." He found, while at his experimental station, that breeding contributed to a 10 percent increase in yield and that soil fertility improvement through organic matter and green manures contributed to a 200 to 300 percent increase.

We find the same at the Navdanya farm. The same variety of crop produces more with fertile soil. "Yield" is not an essential characteristic, fixed and immutable. It is a relational characteristic, a result of its context, especially the soil. It is a potential that thrives with our love for the soil and the seed. And healthy soils produce healthy plants. As Howard stated, "The birthright of every crop is health."

This is especially true in times of climate change. Industrial agriculture is responsible for 40 percent of the greenhouse gases contributing to climate change, and monocrops are also more vulnerable to climate chaos.

During the 2009 drought, when I visited Navdanya members in different parts of the country, I found that their crops had not suffered because they were using locally adapted seeds and their soils had water-holding capacity because of organic manuring. Farmers using Green Revolution varieties or GMO Bt cotton had a crop failure because neither the seed nor the soil were drought-resistant.

Growing diversity and growing organic have become necessary for adapting our soils to climate change.

Supporting healthy soils is the most effective way to get carbon dioxide out of the atmosphere. Soils with organic matter are more resilient to drought and climate extremes. Biodiversity-intensive systems — which are, in effect, photosynthesis-intensive systems — drive carbon dioxide out of the atmosphere, into plants, and then into the soil through the Law of Return.

Soil, not oil, holds the future for humanity. An oil-based, fossil-fuel-intensive, chemical-intensive industrial agriculture unleashes three processes that are killing the soil and hence closing our future.

First, industrial agriculture destroys living soils through monocultures and chemicals.

Second, an oil-based paradigm intensifies fossil fuel inputs and creates a false measure of productivity. Is it "productive" to use ten kilocalories of energy to produce one kilocalorie of food? Why not rely on truly productive, soil-building systems that use one kilocalorie of energy to produce two kilocalories of food?

The trick to getting this crucial math right lies in recognizing that creative, productive human labor and fossil fuels can both be considered "inputs." Intensive fossil fuel use — the bedrock of industrial agriculture — translates into more than three hundred "energy slaves" working invisibly behind each worker on fossil-fuel-intensive industrial farms. In Big Ag, which considers only people as "inputs," the fewer people on the land, the more "productive" agriculture becomes. Farmers are destroyed, rural economies are destroyed, the land is emptied of people and filled with toxics. Deadly chemicals replace the creative work of farmers as custodians and renewers of soil and soil biodiversity.

Let's retire the sad notion of people and land as commodities. Creative work—being stewards of the land and cocreators of living soil —is not an "input" into a food system, but the most important output of good farming. It cannot be reduced to a "commodity." Land, too, is not a commodity. Creating, conserving, rejuvenating fertile and living soil is the most important objective of civilization. It is a regenerative *output*.

Third and last, displaced farmers flood cities. This is not a natural or inevitable phenomenon. It is part of the design of industrial agriculture. It harms both people and land, directly. The explosion of cities buries fertile soil under concrete. The equivalent of thirty football fields are consumed by cement and concrete every minute.

As a lover of soil, and the international patron of the Save Our Soils (sos) movement, I will not rest until humanity wakes to the soil emergency. For too long, there has been an avoidance of life. People moving away from the land. People not wanting to get their hands dirty. Life is what sustains life! The soil is the limit, and what a beautiful, strong, productive, deserving limit it is.

In September of 2012, I joined Klaus Topfler, the head of the UN Environment Program, and Volkert Engelsman, founder of Nature & More, for some guerrilla gardening in the heart of Berlin. It was Soil Week. We removed a section of uniform, gray, flat pavement squares and planted a garden.

On February 12, during the largest organic food trade fair in Nuremburg, I helped with more guerrilla gardening. Sarah Weiner, a celebrity chef, joined me with her shovel. We removed concrete to liberate the soil and sow seeds of freedom.

Living seeds and living soils are the foundation of living and lasting societies.

We will liberate the seed. We will liberate the soil. And through that, we will liberate humanity.

CITY DIRT

❋

KAREN WASHINGTON

rowing food has meant so much to me. Never imagined growing up in a big city like New York that I would feel such passion. Yeah, that's right I am growing food in New York City, da Bronx to be exact. No I don't come from a family of farmers, we're just city folk; but somewhere in my DNA there lies an ancestral lineage to the land, agrarian people, or so I'm told. And I'm a believer.

Let me begin by telling you a little bit about my personal relationship with food. I was born and raised in New York City, as a matter of fact. So were my parents. Neither they nor my grandparents, who came from the South, had any farming experience.

Growing up, food was something my mother cooked, and what a cook she was. She *be* in that kitchen cooking our favorite food, southern fried chicken, mac and cheese, and collard greens with ham hocks or smoked neck bones. Everyone knew when she *threw down* because the aroma would permeate our fourth floor apartment. And make a few neighbors wish they'd been invited. We never questioned who grew the food, what farm it came from, or if it was treated with chemicals. Back in those days all you cared about was how good the food tasted and how much it cost.

As a single parent with two children, I moved to the Bronx in 1985 and purchased my first home. I graduated from college with two degrees, a Bachelors of Science in physical therapy and a Master of Arts degree in biomechanics and ergonomics. Yet with all these degrees and academics, I too had no farming or gardening experience.

I did however have a huge backyard and three options: cement it, cover it with grass, or grow food. I chose to grow food. Now remember back in the day, the Internet was in its infancy, so I relied mostly on books I had purchased, the library, and the words of wisdom from my elders. This, you can say, was my first encounter with growing food and greeting dirt.

I was so excited about what I could grow. The pictures in books and catalogues were so surreal, I wondered if I could ever grow such beautiful vegetables. So I began planting the usual, tomatoes, peppers, eggplant, and collard greens. You know I had to grow collards. You see, growing up, collards were special in my family. We ate them during all major holidays and family gatherings. As part of my African American culture and tradition, they were meant to bring you good luck, and luck is what this city girl was counting on.

As I began growing food, I felt my palate changing. Gone were the strong cravings for salt and sugar. It began with my first bite into a homegrown tomato. Prior to that, I hated tomatoes, stuffed in a plastic carton at the grocery store and tasting like cardboard. So when I took that first bite into the tomato, that I grew, with my hands, plastered with dirt, I felt the juiciness of sunshine tingling throughout my body as warm juices ran down my chin and arms. Making sure not one drop reached the ground, I licked my arms and hands covered with dirt to continue to enjoy the essence of sunshine — it seemed the natural thing to do.

Wow what a taste. I had never experienced taste like that before, and I knew then I would never forget it. I was hooked. So hooked that I wanted to grow everything in sight, from collards to mangoes, but sadly found out that I could not grow tropical food and all the reasons why. Still it kept me exploring and trying out new vegetables, as my palate longed for the nuances of tastes such as bitter, sour, tart, aromatic, peppery hot, and dirt.

You see, when touching the earth, I feel my heart beat and the pulse of past generations. When I take a handful of dirt and smell how sweet it is, it takes me to a place of peacefulness. When I grow food and eat it, hmm . . . it makes me feel good inside, and when I share my food with others it makes me smile.

In 1988, I helped start the first community garden in my neighborhood, the Garden of Happiness. It was a trend seen in most urban areas across the country, from Oakland to Philly, Boston to Detroit.

This was due in part to the fiscal deficits cities were experiencing back then. New York City at one time had more than fifteen thousand vacant lots, mostly in low-income communities of color. Many people started taking over these vacant lots as a way to take back their neighborhoods from despair and drugs, while others found it a hands-on means to grow food and/or beautify their neighborhoods. For me it was the beginning of my involvement in food and social justice.

How wonderful Mother Nature is to have given us this natural resource to feed the world. Yet with all that has been graced amongst us there lies hunger, pollution, and waste.

I sit and contemplate the reasons for such degradation to dirt and wonder why. I'm told that community gardens are the lungs of the city, that there's food sovereignty in what we do. Yet each day, someone, somewhere goes without food.

As I started growing food in the city, I became more aware of the type of food that was in my neighborhood and the role it had on the health and well-being of my neighbors. I live in congressional district 16, one of the poorest districts in the country. Many of the residents are on fixed incomes or unemployed. The median income level for a family of four is $23,000, below the poverty line. Most survive on WIC assistance and food pantries.

As a health care professional, I started to see firsthand an increase in food-related diseases such as diabetes, hypertension, heart disease, and obesity in adults and in children. How quickly we were becoming a supersize nation hooked on fast food, processed food, and cheap food, and I dreaded the long-term impacts it was having on our health.

I hear people call my neighborhood a food desert. But I say, "No, Please don't call it that!" Call it by its true name — "hunger and poverty" or call it "food apartheid."

We have lots of food here! We have restaurants, fast food chains, and bodegas! What we don't have are healthy food *choices*.

So what does this say about our food system?

And what were the questions that needed to be addressed?

How often do we take the time to really think about our food? Many of us don't. Do we make a conscious effort to ask the grocer, "Who was the farmer or farm that produced the food?"

Were pesticides used?

Are we concerned about the true value of his or her crop?

Are farm and restaurant workers being treated and paid fairly?

It's not just about the food we eat, but the food system itself, which encompasses so much more. It just cannot be singled out as one problem, having one solution. Because the further away we are from the land, the more detached we become from our food source. Therefore we don't get a chance to care, and why should we? We have placed that responsibility in the hands of the government *for the people, by the people.* We rely on lobbyists and big business.

In the land of milk and honey, have we become the land of greed and money?

Which brings me back to why I grow food. I grow food because food is a game changer, an equalizer. Food does not discriminate; it sustains us and we all have to have it. In a country with such vast resources, hunger and poverty are unacceptable. Poverty and hunger equal shame, and I cannot turn my back on shame.

I am just one of many realizing in a land of plenty this ought not to be, so I'll keep digging, and digging and digging in dirt, until my hands grow weary from the pain of poverty and hunger.

I see people around the world trying to change things, wanting to return to dirt. They are rolling up their sleeves and getting down and dirty again. They are laying down the plow and picking up the shovel for the right to own land and grow food, not industrial food, but food with a purpose. Food that is healthy and nourishing. These enterprising folks want to care for our environment and its animals and resources. Many have lost their livelihood for this cause. In the past, small farmers throughout the world have lost their land to big agribusiness in the name of progress. Some have even committed suicide, as duly noted in India. The time is now to change things. Food sovereignty is what we are calling it. The right to grow food that is healthy, local, fresh, culturally appropriate, and on land we own.

In the Bronx, there are more than 130 community gardens and several community-run farmers markets, CSAs, and food shares. I helped start a farmers market ten years ago when I was told "my kind" couldn't afford it. Guess what, "my kind" can afford it and love it.

Family farms and community gardens are not only happening in this country but popping up all over the world as people want to go back to the land, back to dirt, and it's our youth who are leading the way. Give a kid a two-inch-tall zucchini plant and a patch of dirt, and watch her change the world.

I will continue to be outspoken in my quest to see a food system that is fair and just. So I leave you with this:

The fight to end hunger and poverty will never succeed from a handout, rather, the power lies in a society where wealth and resources are shared amongst those with less, and those with less are given the opportunity to obtain wealth and resources and are deemed powerful.

It all starts with dirt. And then? Seeds. Water. Sunshine. Willing people. The unstoppable power spreads from there.

SOIL VERSUS OIL — KALE
VERSUS KOCH

❋

ATINA DIFFLEY

*O*ur world spun that late afternoon in April — just in from planting two acres of kale — when I opened the letter marked "Minn-
Can Project." The warm spring sun streaming through the west office
window carried the promise of spring. The piece of paper informed
us that MinnCan, owned by the notorious polluters, Koch Industries,
had filed an application to build a crude oil pipeline. I could smell
fresh earth in my hair. The route map accompanying the letter drew
a bold red line through the just-planted kale. My finger, tinted brown
and green from slipping thousands of plants into the soil, stained the
paper as I traced the brash red line.

I raced out of the house and down to the shed where my husband,
Martin, was setting a cultivator on a 1979 Farmall 140 tractor. I waved
the letter in his face, spilt out the details, and blurted, "You have to
call them up and tell them they can't put it through here."

He straightened up, set the box end wrench on the cultivator arm,
and slipped out of his gloves. He's a musician by night and his hands
are well cared for: smooth, flexible, and quick on the strings. They are
no indication that by day he's a man of the soil with seventy-five acres
planted to organic vegetables and soil-building crops. He pushed his
thick black hair off his forehead before saying, "Slow down. What are
you talking about?"

I launched into a long lecture about what he should tell them:
thirty-three years running Gardens of Eagan farm; eighty thousand
Minnesota food co-op members, many of whom would be affected if
the soil that feeds them were threatened; fertility, pest, and disease
management all based on feeding our soil microbial life. It would take
more than our lifetimes to return the soil to what it was, and since we
wouldn't be knifing anhydrous ammonia into the soil and spraying

pesticides to get a crop after a pipeline installation it would mean no organic vegetables from those fields for a very long time. "No one could be so ignorant as to believe that a crude oil pipeline is a compatible land use with a local organic vegetable farm. Obviously, they don't know we are here. You just need to tell them."

"They put these things where they want. You have the answers. You call them."

I am embarrassed to tell you what I said next: "They will listen to you better—you are a man."

He pulled his gloves back on and said, "This one is yours." I could see his hands shaking as he picked up the wrench and tightened down the cultivator shaft. After what we'd already been through, I'd expected him to fight; not just acquiesce.

Crude oil pipelines and public utility applications? I didn't know anything about that.

I knew about soil health and its impact on human health. I knew that the most succulent melons grow on sandy and well-drained soil where they drop their roots deep into the gravelly subsoil and forage for minerals. I understood that our cells are built of nutrients and minerals that plants receive from soil. And even more profound—that our bodies consist of the same elements as wild animals, plants, insects, and soil microbes. We are all soil. We come from the same earth. Soil is life!

But that didn't have anything to do with crude oil pipelines. Or did it?

I did know what installing a pipeline would look like . . . giant machines forcing their way, trampling flat the kale, broccoli, rye, and hairy vetch. Backhoes digging and beeping and belching diesel smoke. Tearing a hole through the greenness of our lives, ripping a seam into the parent material, the mineral source of this landscape. Topsoil sticking to the bulldozer tracks as they lugged through mud —aggregates crushed, microorganisms starving, water channels collapsed. Heavy equipment leaking toxic substances, soaking into the soil . . . where we once grew food.

And once installed: heavy *black*, viscous tar sands oil flowing through the very heart of our farm, carrying the constant silent threat of blowout.

Sixteen years earlier we had lost Martin's fifth-generation family farm to suburban development. I remembered the commitment I made — it seemed like a lifetime ago — sitting in the Woods Field with Martin and our kids, Maize and Eliza, listening to the first bulldozers flattening trees and the trucks hauling away topsoil to be sold. "We're going to move to a new home and land, and I promise you, no one will ever bulldoze it."

After years of searching, we found new land with a base of mineral-rich prairie loam. Thousands of years in the making but damaged in a few short decades — it had been chemically farmed and was badly compacted. The first three years we didn't cash crop it at all, but grew soil-building plants like hairy vetch, rye, and sorghum-sudangrass and incorporated them into the soil with compost to bring back the life. Now fifteen years later, the soil was loose, moist, and producing high-yield crops with rarely a pest or disease issue. I never guessed I would be challenged to keep my promise, and I still naively believed the pipeline company just couldn't know we were there; Gardens of Eagan, producing three million servings of fresh food each season and training new farmers. MinnCan was acting like our soil and farm were fungible — replaceable by another of like kind — and I knew they weren't. So I called them. Picked up the receiver, dialed . . . and hung up after one ring.

You can't do a darn thing to stop hail or frost. But a crude oil pipeline is not an act of God. I took a deep breath — terror doesn't have to mean run; it can mean act.

I dialed again and asked to speak to the project manager. After twenty minutes of holding and transferring, I was connected to the right-of-way group leader. My stomach was still upside down, but I was a little clearer. I started to explain why crossing Gardens of Eagan would never pass the Criteria for Route Selection. She cut me off before I even got to the word *soil*. "We don't change routes. It doesn't matter if you are organic. If you don't like it, go to the public hearing and tell the judge."

The phone went dead.

This system worked for the pipeline company. Of course they didn't want to change it; intimidation paved their right-of-way.

But it wasn't going to work for us, the soil and the people who ate from this land. I knew then, holding that silent phone, that if we wanted something to change, it was up to us to make it happen.

I went outside to the new kale fields, the best place to find my strength. With the route map in hand, I followed the bold red line MinnCan had so brashly drawn across our land, measuring out and pounding stakes through the seven threatened fields. Then I attached twine: stake to stake through the young kale plants, across the lush nitrogen-sequestering hairy vetch, down into and back out of the grass-covered waterway that also serves as beneficial insect habitat, through the just-planted-yesterday broccoli, down the middle of the rye soil-building crop, and, finally, dissecting the field where yesterday the crew had shoveled compost in long lines to prepare for tomato planting. I looked back at the string running the width of our land and imagined what would be there after the bulldozers had come and gone.

Acres of scraped-raw, compacted dirt. Veins of erosion scouring the surface. Then the tiny fingertips of grass — our attempt to rebuild the soil — but no vegetables growing. That was the exact moment I switched from feeling like a helpless victim to stepping into our role as guardians of the soil.

I went to the online docket to review the application filed by the pipeline company and found something called an Agricultural Impact Mitigation Plan. With this plan, MinnCan (MPL) claimed they could return the soil to pre-pipeline conditions. But any farmer — organic or chemical — knows this isn't possible, so I looked up the word *mitigate* in the dictionary.

Mitigate. (1) To make an offense or crime less serious or more excusable. (2) To make something less harsh, severe, or violent. Etymology: past participle of *mitigare*, to soften.

They had the right word as far as offense or crime. To mitigate doesn't fix or solve the problem, just reduces the sting, softens, excuses. *Excuse me?* They thought they could just bulldoze through here and say, "Excuse me!"

I still thought I might be overreacting until I started reading the plan and came to "MPL will not knowingly allow the amount of top cover to erode more than 12 inches from its original level." And, "MPL may employ temporary, non-destructive uses of topsoil such as creating access ramps at road crossings." *Not knowingly* excused nothing. As if twelve inches of soil loss wasn't enough, they called access ramps a nondestructive use of topsoil. They clearly knew nothing about soil microbial life and aggregation.

Enough of this foolishness. This plan was not what any farmer would classify as an acceptable excuse.

For the next week I ran to the house and called lawyers any time I could break free of spring planting. Each one said the same thing, "I'm sorry we can't help you. We have a conflict of interest." Had Koch tied up every attorney in the Twin Cities with retainers? Finally, environmental attorney Paula Maccabee sang the right tune in my ear. "Gardens of Eagan organic farm, crude oil pipeline, twelve inches of topsoil erosion? This is going to be fun!"

When she came to the farm for our first meeting, I knew I would like her, a woman who laughs at the challenge of taking on the Koch brothers. But I didn't know she would be a buff firecracker, skilled in martial arts, with luxurious, soft black curls that I wanted to reach out and touch. I knew I had the right partner as we established that Gardens of Eagan would intervene in the legal proceeding as a full party to the case. Then I asked Paula what it would take to stop the pipeline altogether. She said, "Just prove in court that society doesn't need the oil. Can you do that?"

"The soil and the farm can do that!" Research from thirty-year trials at Rodale Institute shows that organic farms use 45 percent less energy while sequestering 15 to 28 percent more carbon—with equivalent yields to non-organic farms. If U.S. agriculture changed to organic systems, we wouldn't need the additional oil, and we would reduce agriculture's contribution to greenhouse gas emissions. "We have fifteen years of soil tests, yield, and input records to show the change in the soil and our yields since we transitioned this farm out of chemical management."

But less than 1 percent of U.S. agricultural land is certified organic. Paula said, "We can't show society is ready."

How ironic. Koch Industries is one of the largest manufacturers of synthetic nitrogen and the second largest privately owned company in the world. Here we were, a local organic farm supplying our fertility needs using the renewable energy of the sun and legume plants, threatened by one of the world's largest producers of fossil-fuel-based fertilizer.

Our strategy became clear when Paula explained, "We can provide evidence that organic farms are valuable natural resources that should be protected as such." Hallelujah! She not only had the necessary moxy and chutzpah, she understood the value of organic systems. Under National Organic Program standards, the "physical, hydrological, and biological features of an organic production operation, including soil, water, wetlands, woodlands and wildlife," are defined as natural resources of the operation.

We set our goals: to create an Organic Mitigation Plan that specifically protected the soils and certification of Minnesota organic farms from public utilities, and to move the pipeline off of Gardens of Eagan and other threatened organic farms.

Paula closed our meeting with a practical challenge. "Your job is to provide the education and teach about organic farming. I'll take care of the fight."

And fight she did! For the next months Martin and I planted and cultivated during daylight hours, and I spent most nights working with Paula, writing affidavits and an Organic Mitigation Plan, strategizing "discovery" questions, and working with our expert witnesses. Together we established for the legal record how organic farming systems are based on soil health and why organic farms must be protected.

Paula's advice, that education played a critical role in this battle, came to the forefront as we engaged in the legal process. The Public Utilities Commission (in charge of permitting and routing decisions) and the Minnesota Department of Agriculture (which manages the Agricultural Impact Mitigation Plan) didn't know how organic systems work. That was the point of inspiration for me: citizens can impact policy decisions as educators. I am not an expert on crude oil and public utilities. I am an expert in my own field—organic farming—and that would be the source of my influence.

The Department of Agriculture's land-use employee, Bob Patton, asked excellent questions. "How will the pipeline's impact on organic farms be different than on non-organic?" The best way to answer was to invite him out and let the farm speak for itself. We walked through a dense field of waist-high hairy vetch, capable of sequestering up to 160 pounds per acre of atmospheric nitrogen—our primary source of fertility. We stopped at a rye field that had been disked the day before and found it covered with a complex network of shimmering spider strands—miles of glistening threads, sticky pest insect traps, slung like hammocks—a tribute to healthy soil and biological diversity-based pest management. This was all new for him. Emotion rose as he tasted fresh kale for the first time and said, "I like this!" I swallowed the—*this could be our last crop from this soil*—lump in my throat.

"What should I do with this information?" Bob asked. I handed him our Organic Mitigation Plan.

Paula emphasized that informed-citizen input would be crucial—citizen-educators speaking from their place of experience: how they would be impacted if the pipeline were allowed to pass through Gardens of Eagan and why organic farms should be protected. She explained, "Public clamor is interesting and attracts media attention, but it's not effective in a legal proceeding. Informed-citizen input must be sent to the judge and entered into the legal record for decision makers to use. If we can get two hundred letters, it would really impress the judge."

"Two hundred?" I blurted. "We can get that the first day. People get involved when their personal food source is threatened."

She grinned and flexed her biceps.

Our customers, the Twin Cities natural foods co-ops, got the word out. Letters poured in from university scientists, produce managers, and doctors, from third-generation Gardens of Eagan cornavores and chemically sensitive clients. Dr. Carl Rosen, soil scientist, wrote, "A pipeline would permanently alter the soil temperature, making it impossible to restore the site to a natural condition. In my opinion, constructing a crude oil pipeline through a certified organic farm is an oxymoron."

Wholesale buyer Rick Christianson testified on the loss to organic eaters: "Gardens of Eagan is a source of quality, local-organic food that is irreplaceable in this marketplace."

Letter writers spoke of the importance of preserving organic soil, about peak oil and food security, of renewable fertility systems. They spoke of their family food traditions and multiple generations eating Gardens of Eagan produce. They said, "We have a treasure here that cannot be replaced."

"I wish you could see the care, research, and planning that goes into the raising of each crop, see the diversity of the fields, and the use of the natural watershed in determining the approach to husbanding the land, and taste the sweetness of the corn and the tomatoes and the cabbage that grow," folk musician and poet Bill Hinkley wrote for the legal record.

By the time we were harvesting melons in mid-August, 4,600 letters had been sent to the judge. MinnCan could compensate us — the farmers — for our loss, but how could they compensate 4,600 customers?

They couldn't. And nine months later, after the season's harvest was finished and the soil was planted to its protective winter cover, we had accomplished all our goals.

Organic agricultural land in Minnesota is now recognized as a unique feature of the landscape and treated with the same level of care as other sensitive environmental features. The pipeline was moved to the road right-of-way instead of crossing Gardens of Eagan. The Organic Mitigation Appendix to the Agricultural Impact Mitigation Plan provides protections to the soils and certification of Minnesota organic farms, and it doesn't allow for *any* topsoil erosion! Wisconsin has since adopted a similar organic mitigation plan and activists in others states are working to protect organic soil too.

But something even more powerful was accomplished. The farmer affidavits, expert testimony, and informed-citizen input had educated the judge. She not only had the evidence on record to support our goals, she now understood organic systems and added additional protections of her own.

Koch Industries had deep pockets to use in their fight, but the soil and the farm had the support of the people.

We come from different climates, landscapes, and backgrounds. Whether we eat beans grown on red clay, or tomatoes from brown sand, the minerals build our bodies, the stored energy pulsates life. We each learn about the importance of protecting soil through our unique life experiences, but they all lead us to the same basic truth. *We are all soil. We come from the same earth. Soil is life.*

As a farmer I've learned to read soil in its different forms and work with its seasons. I know in spring that clay will ball into hard clumps if we work it too wet, and we'll be stuck with those clumps the entire growing season — the only thing that will break them is winter's frost. I can squeeze a ball of moist soil and determine rough percentages of its mineral composition: clay, sand, and loam, what its nutrient and water-holding capacity is, and which crops will thrive or fail in it. If our crops have pests or if discolored leaves reveal nutrient deficiencies, I sit barelegged in the field, hands in the soil, feeling and listening until I know what the soil is hungering for. Then we feed it with plants and the renewable energy of the sun rather than reaching for a chemical pesticide or fertilizer.

But I never expected soil would read me, analyze what I am made of, identify my character, and challenge me to give up my victim patterns and act as a soil sister in defense of this most essential element.

The soil is our ancient guide and teacher. Long ago it was naked rock. Time and water, sun and cold, broke the rock into stones, and the stones into dust. For a very long time the earth sat aging; then the life process started and living soil was created.

Eating in the present is a relationship with the past. Our time here is so short, yet our impact is long. Our daily relationship with soil dictates our wellness and determines the future.

Eat, relate, and advocate for the soil that feeds you.

CONTRIBUTORS

PIR ELIAS AMIDON, a student of Sufism for more than forty years, has authored many books on spirituality, including the bestselling anthology *Earth Prayers* and a nondual guidebook titled *The Open Path*. He is the spiritual director of the Sufi Way, a nonsectarian mystical school.

JULENE BAIR is the author of *The Ogallala Road: A Memoir of Love and Reckoning* and *One Degree West: Reflections of a Plainsdaughter*. Her essays have appeared in the *New York Times* and *High Country News*. She lives and writes in Longmont, Colorado.

BOB CANNARD has been farming sustainably for thirty years. Son of a nurseryman, Cannard was among the first to establish farmers markets in northern California in the 1970s. Today he applies his sustainable food intelligence to the bounty of Green String Farm in Petaluma, California. He sells food to Chez Panisse and other beautiful locavores.

FRED CLINE, vintner extraordinaire, founded Cline Cellars in the San Francisco Bay area. Since 1982, he has restored many ancient vine sites to their rightful reigns as premier California vine lands. He pioneered the cultivation of Zinfandel and Rhone-style varietals in areas that had been primarily devoted to Merlot, Cabernet, and Chardonnay grapes. Fred is one of the original Rhone Rangers.

ATINA DIFFLEY is an organic farmer-educator, activist, and author of *Turn Here Sweet Corn: Organic Farming Works*, the winner of the 2013 Minnesota Book Award in memoir and creative nonfiction. Her life's work is supporting sustainable food practices.

DEBORAH KOONS GARCIA'S 2004 film *The Future of Food* offers the first in-depth look into the history and technology of genetic engineering. Her 2012 documentary *Symphony of the Soil* reveals the miraculous power of healthy soils to sustain us. Garcia owns her own production company, Lily Films, which is based in Marin County, California.

EBAN GOODSTEIN, an environmental economist, has published *Economics and the Environment* and *Fighting for Love in the Century of Extinction*. His passion for education created the advocacy group c2c, a college-based forum bent on creating leaders for a new, sane, environmentally balanced future.

BERND HEINRICH, a biologist, has written diverse books that combine scientific insights with striking literary skill. A professor emeritus at the University of Vermont, he is an avid runner who runs with the same intensity he brings to the world of biology. *Ravens in Winter, The Snoring Bird,* and *Why We Run* have won readers throughout the world.

PETER HELLER is the author of the bestselling novel *The Dog Stars*, which was the Apple iTunes Novel of the Year and featured in more than a dozen Best Books of 2012 lists. His novel *The Painter,* was chosen as a Top 20 Amazon Best Books of 2014. Peter lives in Denver and Paonia, Colorado.

LINDA HOGAN, a speaker, poet, playwright, novelist, memoirist, professor, and environmentalist, has won numerous awards for her lyrical work. Her books include *Rounding the Human Corners, People of the Whale,* and *Dwellings, A Spiritual History of the Land.* She is currently working on a book about Chickasaw history, mythology, and lifeways.

PAM HOUSTON has won numerous awards for her delicious take on all things western, including the Pushcart Prize, the O. Henry Award, and the Western States Book Award. Cheryl Strayed calls her newest novel *Contents May Have Shifted* "ravishing." Pam teaches creative writing at UC Davis and sweeps dirt daily from the surfaces of her ranch house in the wilds of Colorado.

WES JACKSON, a natural writer and land advocate, founded the Land Institute to transform agricultural practices nationwide. His books *Nature as Measure* and *The Genius of the Place* are just two of the many powerful books he's written that reflect his stubborn love of dirt.

EDWARD KANZE, an author, naturalist, photographer, and wilderness guide, has published six books. Whether *Kangaroo Dreaming* in Australia or exploring his roots in *Adirondack: Life and Wildlife in the Wild, Wild East,* Ed offers expansive, insightful views into the workings of nature. His warmth is contagious.

JOHN KEEBLE has written five novels, including *Yellowfish* and *Broken Ground.* His riveting eco-thriller *The Shadows of Owls* came out in 2013. Professor emeritus of creative writing and English at Eastern Washington University, he has also published a collection of stories, *Nocturnal America,* and a searing nonfiction expose titled *Out of the Channel: The Exxon Valdez Oil Spill in Prince William Sound.*

LISA KNOPP is the author of five collections of essays. The most recent, *What the River Carries: Encounters with the Mississippi, Missouri, and Platte,* won the 2013 Nebraska Book Award. Knopp is an associate professor of

English at the University of Nebraska Omaha, where she teaches courses in creative nonfiction.

MARILYN KRYSL, a poet, essayist, and fiction writer, has been published in *The Atlantic*, *The Pushcart Prize Anthology*, and in *Best American Short Stories 2000*. "Krysl is funny, fierce, and feminist in the best possible way," said John Updike, "and a technician of variety and resourcefulness."

CHRIS LARSON, a green architect, designs earth-friendly homes. He admits to being hooked on passive solar. He draws inspiration from the behavior of form and pattern in nature. He also teaches and practices Zen Buddhism in North Carolina.

BK LOREN won the Reading the West Book Award as well as the WILLA Award in fiction for her 2012 novel *Theft*. Her work has been nominated for Pushcart Prizes three times. An editor once told her she writes like she was raised by wolves. She tries to live up to that daily.

DAVID R. MONTGOMERY is the author of *Dirt: The Erosion of Civilizations* and *The Rocks Don't Lie: A Geologist Investigates Noah's Flood*. His childhood love of maps led him to study and teach geomorphology at the University of Washington.

ERICA OLSEN loves desert dirt. She has worked as an archivist and museum technician, helping preserve archaeological collections in Utah and Colorado. "Grand Canyon II," one of the short stories in her debut book *Recapture*, won the Barthelme Prize for Short Prose. She hikes, works, and writes in the Four Corners.

JOHN T. PRICE, an advocate for Iowa's Loess Hills, has published the memoirs *Daddy Long Legs: The Natural Education of a Father* and *Man Killed by Pheasant and Other Kinships*. He directs the nonfiction writing program in the English Department at the University of Nebraska Omaha. His newest book, *The Tallgrass Prairie Reader*, honors the diversity of the prairie he calls home.

LAURA PRITCHETT won the WILLA Award for her 2007 novel *Sky Bridge*, and her earlier novel *Hell's Bottom, Colorado* won both the Milkweed National Fiction Prize and the PEN USA Award. Her latest novel *Stars Go Blue* skillfully blends the landscape of old age with the harsh realities of a Colorado winter.

JANISSE RAY won the American Book Award for her first book, *Ecology of a Cracker Childhood*. She has written for *Audubon* and *Orion* and has published several books devoted to Georgia's life and wildlife, and the great Altamaha River. A devoted naturalist and activist, she tries to live a simple, sustainable life on her farm in southern Georgia. Gathering eggs and milking cows, what better way to start a writing day?

BARBARA RICHARDSON has written two novels, *Guest House* and *Tributary*, both of which reflect her ardor for life in the West. *Tributary* won the Utah Book Award and was a WILLA Historical Fiction Finalist. Barbara co-edited the 2015 anthology *I Am with You: Love Letters to Cancer Patients.* She writes and edits in the Wasatch Back.

JANA RICHMAN, writer and desert acolyte, draws strength from southern Utah's red rock canyons. She is the author of a memoir, *Riding in the Shadows of Saints: A Woman's Story of Motorcycling the Mormon Trail,* and two novels, *The Last Cowgirl* and *The Ordinary Truth. The Last Cowgirl* won the 2009 WILLA Award for Contemporary Fiction.

JEANNE ROGERS, a writer and painter, grew up on a Midwest farm. Her memoir, *Changing Course,* chronicles her leap from land to sea while employed as a steward assistant on an oil tanker. In her poetry collection, *Through the Cattails,* she celebrates the interconnectedness of human lives and the places they inhabit.

CARL ROSEN heads the Department of Soil, Water and Climate at University of Minnesota and also teaches in the Horticultural Sciences Department. His publications and extension bulletins on nutrient management, soil fertility, plant nutrition and crop rotation attest to a lifelong dedication to improving and respecting dirt.

DONALD G. SCHUELER has captured the beauty and complexity of the South in two of his seven books, *Preserving the Pascagoula* and *A Handmade Wilderness: How an Unlikely Pair Saved the Least Worst Land in Mississippi.* A former professor of English at the University of New Orleans, his writing embodies environmental empathy and love of place.

DR. VANDANA SHIVA, a philosopher and environmental activist, has published many books, among them *Soil Not Oil, Stolen Harvest, Earth Democracy,* and *Staying Alive.* She is the founder/director of Navdanya Research Foundation, dedicated to the preservation of native seeds, organic farming, and fair trade. She advocates the protection of native soils worldwide.

KAYANN SHORT writes, farms, and teaches at Stonebridge Farm, her organic CSA farm in the Rocky Mountain foothills of Colorado. Her book *A Bushel's Worth: An Ecobiography* unites generations of family farms with her engaging call to action for local farmland preservation.

LIZ STEPHENS' memoir *The Days Are Gods* seeks threads of community in a strange new land. She has won the Western Literature Association's

Frederick Manfred Award and was a finalist for the Annie Dillard Creative Nonfiction Award. Her work has been published in *Fourth Genre*, *Brevity*, *Western American Literature*, and *South Dakota Review*.

ROXANNE SWENTZELL has worked with clay since she was four. Coming from a family of renowned potters and sculptors, she attended the Institute for American Indian Arts in Santa Fe and the Portland Museum Art School. Her skillful, subtle, often playful clay figures—seen in museums and galleries throughout the United States—quiet the mind and enliven the heart.

CARRIE VISINTAINER is a Colorado-based writer. Her work has appeared in *The Huffington Post*, *Outside Online*, *5280*, *Proto*, *Ars Medica*, and various *Travelers' Tales* "The Best Women's Travel Writing" volumes. She received an MS in molecular biology and genetics from the University of Minnesota.

TYLER VOLK, a biologist, has written six books, including *CO_2 Rising*, *Metapatterns*, *Gaia's Body*, and *Death & Sex*, with Dorion Sagan. A brilliant writer and professor, Volk plays lead guitar for the Amygdaloids and lives to explore and explain the complex systems of our living earth.

KAREN WASHINGTON, a community activist, was awarded the James Beard Foundation Leadership Award in 2014 for her inspired work as an urban farmer. She sits on the board of the New York Botanical Gardens and has spent years turning empty Bronx neighborhood lots into verdant gardens. She cofounded BUGS (Black Urban Growers) and La Familia Verde Farmers Market, bringing fresh vegetables to city dwellers.

TOM WESSELS, a terrestrial ecologist, has taught and published with passion for the past thirty-five years. A professor at Antioch University New England, Wessels founded the conservation biology program in the Department of Environmental Studies. His books *Untamed Vermont*, *The Myth of Progress*, and *Forest Forensics* reveal his deep connection to New England and the planet.

CREDITS